0.1 Mechanical Engineering Science

O.1 Mechanical Engineering Science

In SI Units

J. L. Gwyther, B.Sc.(Eng.), B.Sc.(Econ.), C.Eng., F.I.Mech.E.
Vice-Principal, College of Technology, Letchworth

W. D. Brown, M.Sc., C.Eng., M.I.Mech.E., M.T.I.
Sometime Principal Scientific Officer, Ministry of Aircraft Production

G. Williams, B.Sc.(Eng.), C.Eng., M.I.C.E., M.I.Struct.E.
Deputy Engineer, Severn River Authority

H·E·B

Heinemann Educational Books Ltd
London

Heinemann Educational Books Ltd

LONDON EDINBURGH MELBOURNE TORONTO
JOHANNESBURG AUCKLAND SINGAPORE
HONG KONG IBADAN NAIROBI NEW DELHI

ISBN 0435 71473 2

First Published 1970

Published by Heinemann Educational Books Ltd
48 Charles Street, London W1X 8AH
Printed in Great Britain at the Pitman Press, Bath

Preface

The primary objective of the authors in writing this book was to meet the demand for a new lead-in to the engineering science subjects of the Ordinary and Higher National Certificates in engineering in their new forms: but the book will also be found useful for a variety of other courses in engineering science. The ever-increasing need for scientific and technical workers is having its effect on the curricula of secondary schools, and the book will be useful in all courses of post-'O' level or post-C.S.E. standard in the schools. Also, its use in the colleges will not be confined to national certificate courses, for its comprehensive approach and breadth of coverage will make it a valuable reference work in courses in mechanical engineering science for Higher Technicians.

The book is being produced at a time of changeover of units to the metric system. Although this changeover will not be complete until the year 1975 at earliest, it has been deemed essential to plan the presentation to meet interim needs. This has meant the presentation of all dynamical theory in both sets of units, as well as the giving of due regard to the likely change of emphasis in the future from gravitational to absolute units in the SI system. Statical calculations are also presented in both systems, although it is appreciated that the change in practice here will be quicker. The units of the thermodynamics sections have also been presented with the changeover in mind. SI units are indicated throughout.

The syllabus coverage is adequate to meet the needs of all the examining boards. Thermodynamics and Heat Engines occupy a larger place than has been usual in books dealing with mechanical engineering science at this level, to meet the needs of the new schemes of work. Students are advised to look at past papers of the examination they are taking so as to judge the emphasis of the work.

Acknowledgements are gratefully made to the several examining boards which have allowed the use of their questions here. They include the Union of Lancashire and Cheshire Institutes (U.L.C.I.), The Union of Educational Institutions (U.E.I.), The East Midlands Examining Union (E.M.E.U.), and The Northern Counties Technical Examinations Council (N.C.T.E.C.). The book has gained much from these examples.

Messrs Dobbie McInnes have provided the copy for our artist to draw the diagram of the engine indicator, and their help is much appreciated.

Contents

1

Introduction

Mechanical Engineering Science covers a wide range of topics which will be developed later into Applied Mechanics and Heat Engines at O2 level. These subjects may be developed still further at advanced levels.

If you have reached your present grade by taking the G1 and G2 courses, you will already have been introduced to some of the work of this book, and certainly most of the symbols and abbreviations which are used, and which have been recommended for use in all the engineering courses.

Inevitably there will be some overlapping of material, because further developments of particular aspects may require a greater knowledge of mathematics, which is only acquired in your next year. The amount of revision of your previous work has, however, been reduced to a minimum, firstly because there will be too little time for you to go back far, and secondly because a widening of the subject matter must enlarge its content, and so increase its cost to you.

Aims and scope of this book

The aims in writing this book have been to:

1. Revise, where necessary, and as briefly as possible, work which has been introduced at previous levels.

2. Expand this material, so that you are able to tackle more difficult problems.

3. Emphasise certain parts of the syllabus which are considered important by the various examining bodies.

4. Present new work in a simple manner so that you can understand it fully.

5. Introduce work which will be explained in greater detail in the next year of the course, when you will have acquired a greater knowledge of other subjects, particularly mathematics.

6. Encourage you to carry out every experiment intelligently, and to write up your work in a proper manner.

Worked examples, which have been selected with great care to cover almost every type of question you are likely to meet in an examination, have been introduced into the text. They are solved simply and in great detail. Further questions, with answers, can be found at the end of each chapter. Many of these questions are from back papers of examining boards. All diagrams bear a prefix number which is the same as the number of the chapter with which they are associated. This simplifies reference from one to another.

2

The equations of motion

The fundamental dimensions of mechanics are *mass*, *length*, and *time*. In this chapter we shall be concerned only with the last two. In mechanics length is usually specified in *feet* in the f.p.s. system of units; or in *metres* in the SI (Systeme International) system. Time is specified in seconds in both systems. There are, of course, other units of length and time, but they must be converted for the purpose of dynamical calculations. Certain other terms, most of which we have already met in earlier courses, are defined below.

Vector
A vector is a straight line representing to scale both the magnitude and direction of some physical quantity. If the direction of the quantity is related to some fixed or known direction, then the vector provides sufficient information to enable everyone to define the quantity exactly.

Displacement (symbol s)
Displacement represents the distance moved by a body from its starting position in a definite direction. It is therefore a vector quantity, and represents motion in a straight line. For the purposes of mechanics, displacement should be specified in feet or in metres.

Velocity (symbol v)
Velocity is the rate of change of displacement; in other words, the change (increase or decrease) of displacement in unit time. This also is a vector quantity, and represents motion in a straight line.

If the amounts of displacement in successive units of time are constant, and take place in the same direction, then the velocity is constant, and we can say

$$v = \frac{s}{t}$$

The units of velocity will be feet per second (ft/s), or, in the SI system, metres per second (m/s).

When the amount of displacement in unit time varies, or the direction changes, then the velocity is not constant. But if we consider a unit of time small enough, we can determine an *instantaneous velocity*, which is the velocity at a particular instant. An instantaneous velocity is the amount of displacement which would have occurred in, say, the next second had the body continued to travel in the same direction without any change in its motion.

Distance (symbol s)
Whenever we are not concerned with straight line motion, we do not use the word 'displacement'. Instead we use 'distance'. This is the actual length of the path taken by the moving body from one point to the next. Distance is not usually the subject of dynamical calculations, so we can expect to find it recorded in any length units such as inches, feet, yards, miles, metres, kilometres, etc.

If the motion happens to be in a straight line, then distance will be numerically equal to displacement in that direction if the same units are used, but usually it is not necessary to specify the direction in which distance is measured.

Speed (symbol v)
Frequently we wish to refer to motion which is taking place along a route which is not a straight line. The rate at which a distance along such a route is travelled (meaning distance travelled in a unit of time) is known as the speed. If the speed is measured by a meter in a motor-car it will probably be given in miles, or kilometres, per hour. This rate is an instantaneous speed because it shows the distance which would have been travelled in one hour if the car continued at constant speed.

If, however, a total of t hours were taken by the car to travel a distance s, then the average speed throughout the whole journey would be given by the expression

$$\text{Average } v = s/t$$

Obviously, if a *constant* speed were maintained throughout the journey, the uniform or constant speed would be given by

$$\text{Constant } v = s/t$$

If the motion is in a straight line, and the speed is constant, then speed and velocity are the same.

Acceleration (symbol a)
Acceleration is the rate of change of speed or velocity. In this book we shall be concerned only with constant acceleration, meaning constant increase or decrease of velocity in successive units of time.

When there is an increase in velocity in successive units of time, the body is said to be accelerating: if there is a decrease in velocity the body is said to be retarding. Retardation is the same as negative acceleration.

Acceleration and retardation are usually specified in feet per second per second, or, on the SI system, metres per second per second, and abbreviated ft/s^2 and m/s^2 respectively.

Speed conversions

For a variety of reasons we are not often given speeds in 'feet per second'. It is much more usual to find the speed of a train or a car given in 'miles per hour', and that of a ship or aircraft given in 'knots'. Note that we do not refer to time when quoting a speed in 'knots'. This is because 'one knot' is another way of saying 'one nautical mile per hour'.

If we are going to use the equations of motion, we must be able to convert miles per hour, and knots, to feet per second. We can, of course, convert miles into feet and hours into seconds, thus

$$\frac{1 \text{ mile}}{1 \text{ hour}} = \frac{5{,}280 \text{ feet}}{3{,}600 \text{ seconds}} = 1{\cdot}46 \text{ ft/s}$$

But it is much more convenient to factorize, as follows:

$$\frac{1 \text{ mile}}{1 \text{ hour}} = \frac{5{,}280 \text{ ft}}{3{,}600 \text{ s}} = \frac{88 \text{ ft}}{60 \text{ s}}$$

and to say therefore, that

$$60 \text{ mile/h} = 88 \text{ ft/s}$$

Similarly,

$$1 \text{ nautical mile} = 2{,}027{\cdot}3 \text{ yd/h} = \frac{2{,}027{\cdot}3}{1{,}760} \text{ mile/h}$$

$$= 1{\cdot}15 \text{ mile/h}$$

or,

$$1 \text{ knot} = 1{\cdot}15 \text{ mile/h} = 1{\cdot}69 \text{ ft/s}$$

The equations of motion

We have seen that motion involves the use of four physical quantities, namely, distance, time, speed and acceleration. These four quantities can be related to one another by certain expressions which are known as the 'equations of motion'. The same equations may be used to relate the vector quantities displacement, velocity and acceleration, but then we must confine their use to linear motion, which means motion in a straight line. Let

s = distance travelled from the start to some new position
t = time taken to travel distance s
v_1 = instantaneous speed at starting point
v_2 = instantaneous speed at new position
a = constant acceleration during this motion

We will not, in this book, consider any problem involving an acceleration which is varying, but sometimes we might wish to consider a retardation, which, as we have seen, is a negative acceleration.

When the acceleration is constant we can determine the speed (or velocity) at any time after the start, because by our definition of acceleration the speed increases by a every second. Thus, after

1 second the speed is $(v_1 + a)$
2 seconds the speed is $(v_1 + 2a)$
3 seconds the speed is $(v_1 + 3a)$

Therefore after t seconds the instantaneous speed is given by

$$v_2 = v_1 + at \qquad 2.1$$

This is our first equation of motion. Note that it contains four quantities and only four.

We have already said that

$$\text{Average speed} = s/t$$

but,

$$\text{Average speed} = \frac{v_1 + v_2}{2}$$

so that

$$\frac{v_1 + v_2}{2} = \frac{s}{t}$$

or, transposing,

$$s = \tfrac{1}{2}(v_1 + v_2)t \qquad 2.2$$

This is our second equation of motion. Note again that it contains four quantities and only four.

Now the first equation can be re-written in the form

$$v_2 - v_1 = at$$

and the second in the form

$$v_2 + v_1 = \frac{2s}{t}$$

if we multiply these together, we get

$$(v_2 + v_1)(v_2 - v_1) = 2\frac{s}{t} \times at$$

or,

$$v_2^2 - v_1^2 = 2as \qquad 2.3$$

This gives us a third equation, again containing four quantities and only four.

We can eliminate v_2 from our first two equations, by subtracting

$$v_2 - v_1 = at$$

from

$$v_2 + v_1 = 2\frac{s}{t}$$

We should get

$$2v_1 = 2\frac{s}{t} - at$$

or, multiplying both sides by t,

$$2v_1t = 2s - at^2$$

transposing,

$$2s = 2v_1t + at^2$$

or,

$$s = v_1t + \tfrac{1}{2}at^2 \qquad 2.4$$

This gives us our fourth equation, again containing four quantities and only four.

Note that no two equations contain the same four quantities.

Speed-time graph

When the acceleration is constant, the relationship between speed and time can often be shown to advantage in the form of a graph. Suppose that figure 2.1 represents in graphical form the speed of a train, already

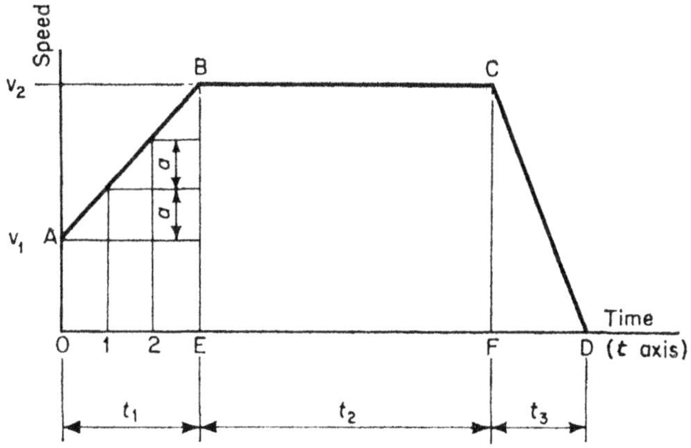

Figure 2.1 Speed-time graph of motion of a train with uniform acceleration

in motion and accelerating uniformly from station A towards station B. If the acceleration a is constant, the speed will increase by equal amounts, each equal to this acceleration, every second. The graph between A and B will therefore be a straight line inclined upwards.

The inclination, or slope, of a speed-time graph at any point represents the acceleration there.

Suppose now that the train, as soon as it reaches station B, continues to the next station C without acceleration—that is, at a constant speed. The graph between B and C must then be a straight line, parallel to the t-axis.

If, finally the train is retarded uniformly until it comes to a halt at station D, the graph between C and D must be a straight line inclined downwards and meeting the t-axis at D, that is, at zero speed.

We have seen, from equation 2.2, that

$$s = \left(\frac{v_1 + v_2}{2}\right) . t$$

or, Distance travelled = Average speed × Time taken

Reverting to the speed-time graph,

1. *Between* A *and* B.

Average speed × Time from A to B = Area OABE

or, from equation 2.2,

Area OABE = Distance travelled from A to B

2. *Between* B *and* C.

Average speed × Time from B to C = Area BCFE

and so

Area BCFE = Distance travelled from B to C

3. *Between* C *and* D.

Average speed × Time from C to D = Area CDF

and

Area CDF = Distance travelled from C to D

That is, *the area under the speed–time graph is equal to the distance travelled.*

This is a most important relationship which may not be immediately obvious from equation 2.2.

Example

A train travels between two stations A and D, 15 km apart, in 24 minutes. The train accelerates uniformly from rest at A, then travels at constant speed for 12 minutes, finally retarding uniformly to stop at D. Determine the value of the uniform velocity in miles per hour. If the time taken for the acceleration is twice the time for the deceleration, find the time taken in each case. (U.L.C.I.)

Solution

Draw a speed–time graph as shown in figure 2.1(a).

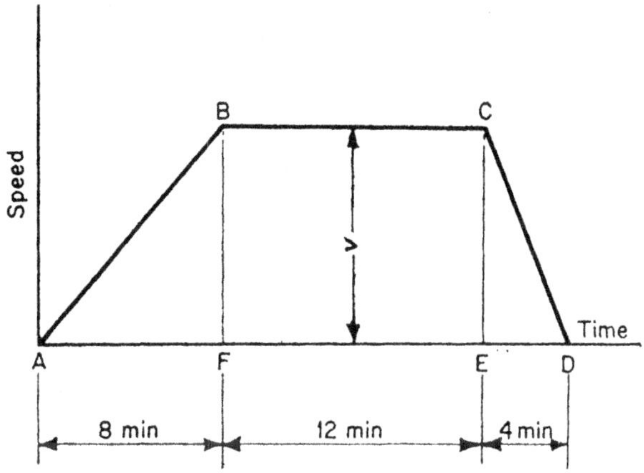

Figure 2.1 (a)

If the constant speed occupies 12 minutes, then acceleration and retardation occupy (24 − 12), i.e. 12 minutes. But acceleration takes twice as long as deceleration. So acceleration takes 8 minutes, and deceleration 4 minutes. Use the figures to draw figure 2.1(a).

The area under the speed–time graph represents the distance travelled, i.e.

$$\text{ABF} + \text{BCEF} + \text{CDE} = 15 \text{ km} \qquad \text{AF} = \frac{8}{60} \text{ hours}$$

$$\text{FE} = \frac{12}{60} \text{ hours}$$

$$\frac{1}{2} \cdot v \cdot \frac{8}{60} + v \cdot \frac{12}{60} + \frac{1}{2} \cdot v \cdot \frac{4}{60} = 15 \text{ km} \qquad \text{ED} = \frac{4}{60} \text{ hours}$$

$$\frac{v}{15} + \frac{v}{5} + \frac{v}{30} = 15$$

$$2v + 6v + v = 15 \times 30$$

$$9v = 450$$

$$v = 50 \text{ km/h}$$

Vertical motion

Suppose that we are asked to consider the motion of a body in a vertical direction, and that we can assume that wind and air resistance are negligible. Although such an assumption is not strictly true, it enables us to understand more clearly how the force of gravity affects vertical motion. At low speeds the assumption does not lead to large errors.

All bodies are attracted towards each other; and the earth, being very great compared with any body on it, pulls everything towards its centre. If, therefore, a body is dropped from a point above the Earth's surface it will fall vertically (that is, towards the centre of the earth) and accelerate at a rate which, for motion at the earth's surface, can be regarded as constant. This gravitational acceleration, or g as it is called, has a value of about 32 ft/s^2, or 9·81 m/s^2. In London it is nearly 32·2 ft/s^2.

We can apply this value of g to the four equations of motion, provided we regard all our quantities as vectors, and take into account their signs.

Convention

Let us assume, as we do when we plot a graph, that all measurements made upwards from the earth's surface, or from some other datum line, are positive. Then all downward measurements must be negative. Let us also assume that time is positive.

Free fall

This deals with the simple case of a body falling vertically from rest, say, from the top of a tower.

In accordance with our convention,

Initial velocity $= v_1 = 0$
Final velocity $= v_2$ downwards, and therefore NEGATIVE
Displacement $= s$ downwards from the start, and therefore NEGATIVE
Acceleration $= g$ downwards, and therefore NEGATIVE
Time $= t$ always POSITIVE

Example

A stone is dropped from the top of a tower 20 m high. Determine:

(*a*) the time of fall;
(*b*) its velocity on striking the ground at the foot of the tower.

Take $$g = 9{\cdot}8 \text{ m/s}^2$$

Solution

If the stone is dropped, and not thrown, its initial velocity v_1 must be zero. The displacement measured downwards, is 20 m and so in the equations of motion we must put $a = -9{\cdot}8$ m/s^2.

It is a useful habit to write down on the right hand side of the page what quantities we know, and what we wish to determine. All the quantities must be in the correct units, and these units should be specified against the numerical values. We shall note for the last time, in working this example, that there must be four quantities, and only four. This will help us to discover the equation of motion that we require, for there are no two equations containing the same quantities.

(*a*) *Time of fall.* The only equation containing the four quantities is

$$s = v_1 t + \tfrac{1}{2}at^2$$

$$(s = -20 \text{ m};\ v_1 = 0;\ a = -9{\cdot}8 \text{ m/s}^2;\ t = ?)$$

Substituting the numerical values, we get

$$-20 = 0 - \tfrac{1}{2} \times 9{\cdot}8t^2$$

$$-20 = -4{\cdot}9t^2$$

$$t^2 = \frac{20}{4{\cdot}9} = 4{\cdot}08$$

$$t = 2{\cdot}02 \text{ seconds}$$

(*b*) *Velocity on striking the ground.* The only equation containing the four quantities is

$$v_2 = v_1 + at$$

$$(v_1 = 0;\ v_2 = ?;\ a = -9{\cdot}8 \text{ m/s}^2;\ t = 2{\cdot}02 \text{ s})$$

Substituting the numerical values, we get

$$v_2 = 0 - 9{\cdot}8 \times 2{\cdot}02$$

$$= -19{\cdot}78 \text{ m/s}$$

The negative sign means, in accordance with our convention, that the final velocity is downwards.

Vertical projection

Suppose that a body is projected vertically upwards from the top of a tower. Owing to the effect of the earth's gravitational field the body will be influenced by a downward acceleration, which, because of our convention, must be regarded as a negative acceleration, or a retardation, upwards. The body will therefore slow down until it reaches the maximum height, when its vertical velocity will be zero. It will then change its direction, and gather speed as it falls. When the body is between A and B (figure 2.2),

Initial velocity of projection
= v_1 upwards, and therefore POSITIVE

Final velocity = v_2 upwards, and therefore POSITIVE

Displacement from the datum line
= s upwards, and therefore POSITIVE

Acceleration = g downwards, and therefore NEGATIVE

Time = t, always POSITIVE

When the body is between B and C.
(*Note:* There is no need to consider what happens between A and B.)

Initial velocity of projection

$= v_1$ upwards, and therefore POSITIVE

Final velocity $= v_2$ downwards, and therefore NEGATIVE

Displacement from the datum line

$= s$ upwards, and therefore POSITIVE

Acceleration $= g$ downwards, and therefore NEGATIVE

Time $= t$ always POSITIVE

When the body is between C and D.

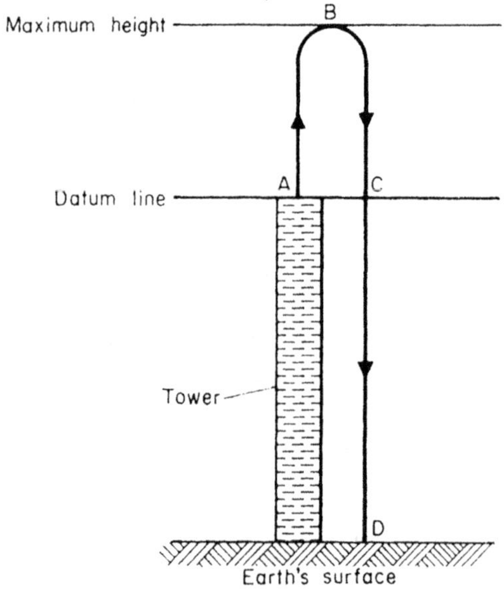

Figure 2.2 Vertical projection

(*Note:* There is no need to consider what happens between A and C.)

Initial velocity of projection

$= v_1$ upwards, and therefore POSITIVE

Final velocity $= v_2$ downwards, and therefore NEGATIVE

Displacement from the datum line

$= s$ downwards, and therefore NEGATIVE

Acceleration $= g$ downwards, and therefore NEGATIVE

Time $= t$, always POSITIVE

Example

A ball is projected vertically upwards from the top of a tower 200 ft high with a velocity of 120 ft/s. Find the maximum height reached by the ball, and also the total time taken from projection for the ball to reach ground level.

Solution

(*a*) Refer to figure 2.2 and consider the path from A to B.

The equation of motion must be

$$v_2^2 - v_1^2 = 2as$$

$$(v_1 = 120 \text{ ft/s};\ v_2 = 0;\ a = -32 \text{ ft/s}^2;\ s = ?)$$

Substituting numerical values,

$$0 - (120)^2 = -2 \times 32s$$

$$s = \frac{(120)^2}{64}$$

$$s = +225 \text{ ft}$$

(The positive sign indicates that the displacement is measured upwards from the datum line.)

(*b*) Refer to figure 2.2 and consider the path from C to D.

The equation of motion must be

$$s = v_1t + \tfrac{1}{2}at^2$$

$$(v_1 = 120 \text{ ft/s};\ a = -32 \text{ ft/s}^2;\ s = -200 \text{ ft (downwards)};\ t = ?)$$

Substituting numerical values,

$$-200 = 120t - \tfrac{1}{2} \times 32t^2$$

or,

$$25 = -15t + 2t^2$$

$$t = \frac{15}{4} \pm \sqrt{\frac{225 + 200}{16}} = 8{\cdot}9 \text{ seconds}$$

(The positive sign only of t is used, in accordance with our convention.)

Horizontal projection

If we are going to assume that wind and air resistance can be neglected, then the velocity in a horizontal direction (or the horizontal component of velocity) must be unchanged. However, the earth's gravitational field will affect the motion, and cause the trajectory to bend downwards. Consequently, horizontal and vertical velocities (or rather, the components of velocity) will be considered separately.

(*a*) *Horizontal velocity* (symbol v_h)

The horizontal component of velocity remains unchanged, and so horizontal displacement (s_h) is determined as follows:

$$s_h = v_h \times t$$

where t is the time to reach a point in the trajectory.

(*b*) *Vertical velocity* (symbol v_v)
The vertical component of velocity is found as in a free fall. Thus, by use of equation 2.1,

$$v_2 = v_1 + at$$
$$-v_{v2} = 0 - gt$$
and
$$v_{v2} = gt$$

This gives the magnitude of the velocity at the point where

$$t = \frac{s_h}{v_h}$$

or, by using equation 2.4,

$$s = v_1 t + \tfrac{1}{2} . at^2$$
$$-s_v = 0 - \tfrac{1}{2} . gt^2$$
and
$$s_v = \tfrac{1}{2} . gt^2$$

This gives the magnitude of the downward displacement where

$$t = \frac{s_h}{v_h}$$

Example

A jet of water issues from an orifice in the vertical side of a tank. The jet is initially horizontal, but it falls through a vertical distance of 16 cm while travelling 160 cm horizontally. What is the velocity of the jet as it leaves the tank?

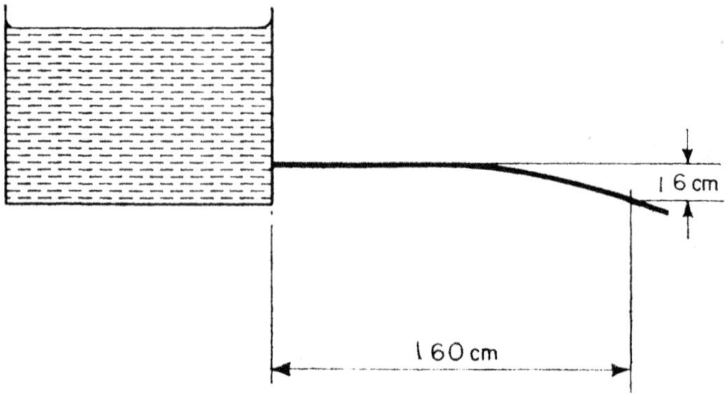

Figure 2.3 Horizontal jet of water

Solution
Let horizontal velocity of water on leaving the tank be v m/s, and time to travel 160 cm horizontally be t seconds.

(*a*) *Horizontal motion*

$$s_h = vt \qquad s_h = 1{\cdot}6 \text{ m}$$
$$vt = 1{\cdot}6 \text{ m}$$

(*b*) *Vertical motion*

$$s_v = v_v t + \tfrac{1}{2} \,.\, at^2 \qquad v_v = 0$$
$$-0{\cdot}16 = 0 - 4{\cdot}9t^2 \qquad a = -9{\cdot}8 \text{ m/s}^2$$
$$t^2 = \frac{0{\cdot}16}{4{\cdot}9} \qquad s_v = -0{\cdot}16 \text{ m}$$
$$t = 0{\cdot}181 \text{ s}$$

and, because $vt = 1{\cdot}6$ m

$$v = \frac{1{\cdot}6}{0{\cdot}181}$$

and Jet velocity $= 8{\cdot}84$ m/s

Angular motion

Many of the problems which confront an engineer are in some way concerned with the rotation of shafts or flywheels. Now the speed of a point on the rim of a flywheel depends on the number of revolutions made by the flywheel in a given time, and also the distance of the rim from the centre of rotation—in other words, the radius of the flywheel.

We usually specify the speed of rotation in revolutions per minute, (written rev/min), and the length of the circumference of the flywheel is determined as π multiplied by the diameter.

The radian

Supposing we take a circle with centre O and radius r, and let OA be a radius rotating in a clockwise direction.

If A moves to B so that the length of the arc AB (figure 2.4) is equal

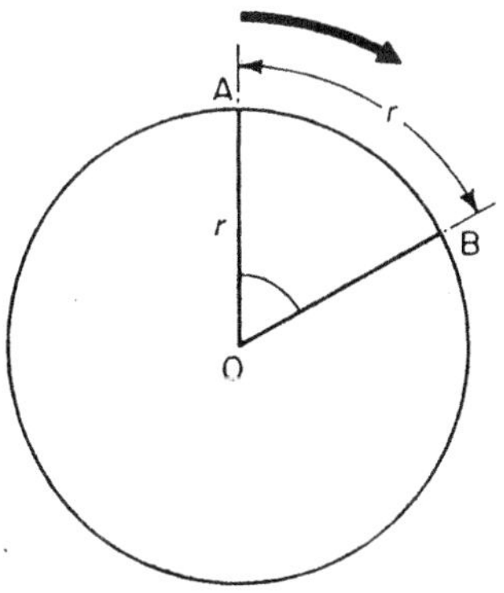

Figure 2.4 A radian

to the length of the radius, we shall say that OA has rotated through an angle of 1 radian. Since the circumference of the circle is given by $2\pi r$, it will be necessary for OA to rotate through $2 \times \frac{22}{7}$, i.e. $\frac{44}{7}$ of these radians in order to complete the circle and return to its original position. A radian therefore, is one way of measuring the angle AOB; another way would be to measure it in degrees.

If the radius OA rotates through a complete circle it will turn through 360 degrees or 2π radians.

That is, 2π radians is the same angle as 360 degrees. Hence

$$1 \text{ radian} = \frac{360}{2\pi} \text{ degrees}$$

or, $$57{\cdot}3 \text{ degrees approximately} \qquad 2{\cdot}5$$

When considering the rotation of a point on the circumference of a circle, we shall be concerned with four quantities, similar to those we used in linear motion. These quantities are angular displacement, angular velocity, angular acceleration, and time.

Angular displacement (symbol θ—Greek letter theta)

The displacement of a point on the circumference of a rotating shaft or flywheel may be determined in terms of lengths of arc equal to the radius. Each such length subtends an angle of one radian at the centre of rotation, and in one revolution the point becomes displaced 2π radians. If, therefore, in time t the point has moved from A to B (figure 2.5)

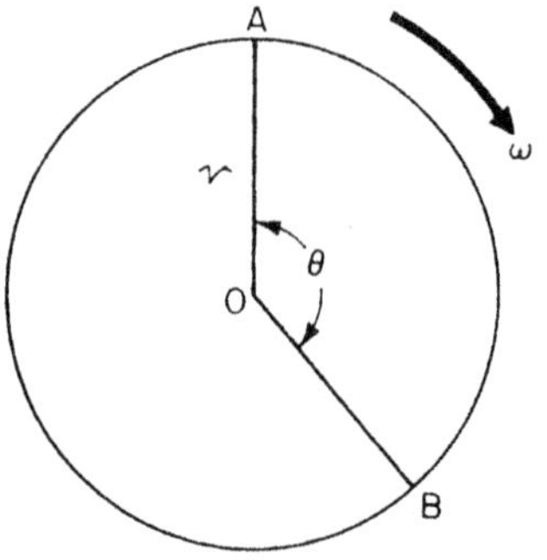

Figure 2.5(a) Angular motion

the angle θ subtended by the arc AB at the centre of the circle can be obtained from the expression

$$\text{arc AB} = r\theta$$

or, $$\theta = \frac{\text{arc AB}}{r} \text{ radians}$$

Angular velocity (symbol ω—Greek letter omega)

If the radius OA (figure 2.5) rotates through an angle of θ radians in t seconds its average velocity is said to be θ/t radians per second.
Hence, if ω is the angular velocity:

$$\omega = \frac{\theta}{t} \text{ rad/s (abbreviation for radians per second)}$$

or, $$\theta = \omega t \text{ rad (abbreviation for radians)} \qquad 2.6$$

If the angular velocity is constant in magnitude, then the radius OA sweeps through equal angles in equal intervals of time.

Angular acceleration (symbol α—Greek letter alpha)

If the angular velocity is not constant, then the radius OA must be accelerating or retarding. Angular acceleration is the rate of change of angular velocity—that is the change of angular velocity in a unit of time.

Relation between rev/min and rad/sec
It is usual to measure the speed of rotation of a shaft in terms of revolutions per minute. This can easily be converted into radians per second. Thus,

$$1 \text{ rev/min} = 2\pi \text{ rad/min}$$

$$= \frac{2\pi}{60} \text{ rad/s}$$

$$n \text{ rev/min} = \frac{2\pi n}{60} \text{ rad/s} \qquad 2.7$$

$$= \frac{\pi n}{30} \text{ rad/s}$$

Conversely, $$1 \text{ rad/s} = \frac{30}{\pi} \text{ rev/min}$$

or, $$\omega \text{ rad/s} = \frac{30\omega}{\pi} \text{ rev/min}$$

Relation between angular and linear velocity
Consider a flywheel of radius r ft, with a point A on its circumference. The flywheel is rotating in a clockwise direction with a constant angular velocity of ω rad/s.
In t seconds OA will have turned through ωt radians to a position OB.

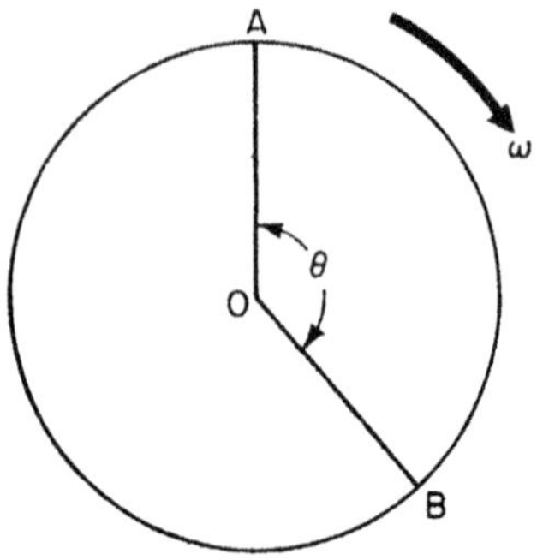

Figure 2.5(*b*) Angular motion

But, Length of arc AB $= \omega t \times r$

where $\theta = \omega t$

Hence Linear speed of A $= \dfrac{\text{arc AB}}{t}$

$$= \frac{\omega t r}{t}$$

or, $v = \omega r$ ft/s 2.8

At any instant the linear velocity of the point A is tangential to the rim of the flywheel, and therefore, although its magnitude is equal to the linear speed, its direction is continuously changing. If

$v =$ Linear velocity

then $v = \omega r$ 2.9

and $\omega = \dfrac{v}{r}$ rad/s 2.10

Relation between angular and linear acceleration

Because acceleration is defined as the rate of change of velocity, whether velocity is linear or angular, we can say, from the above expression that

Linear acceleration = Angular acceleration × radius

So that $a = \alpha r$ 2.11

or, $\alpha = \dfrac{a}{r}$ rad/s^2 2.12

The equations of angular motion

Let suffix (1) denote the initial state, and suffix (2) the state after a time t seconds.

Then, if the initial angular velocity is ω_1 and the angular velocity is increased by α rad/s every second,

$$\omega_2 = \omega_1 + \alpha t \qquad 2.13$$

If the angular velocity is increasing uniformly, that is, the angular acceleration is constant, then

$$\text{Displacement} = \text{Average velocity} \times \text{Time}$$

or,

$$\theta = \left(\frac{\omega_1 + \omega_2}{2}\right) t \qquad 2.14$$

Equations 2.13 and 2.14 can be re-arranged, thus

$$\omega_2 - \omega_1 = \alpha t$$

$$\omega_2 + \omega_1 = \frac{2\theta}{t}$$

and, if we multiply the first by the second, we get

$$\omega_2{}^2 - \omega_1{}^2 = 2\alpha\theta \qquad 2.15$$

If, on the other hand, we were to subtract the first from the second, we should have

$$2\omega_1 = \frac{2\theta}{t} - \alpha t$$

now, multiplying both sides by t, we get

$$2\omega_1 t = 2\theta - \alpha t^2$$

We can divide both sides by 2 and re-arrange to get

$$\theta = \omega_1 t + \tfrac{1}{2}\alpha t^2 \qquad 2.16$$

Note the similarity between these four equations for angular motion and the corresponding equations for linear motion. They are repeated here for easy comparison.

Linear motion	Angular motion
$v_2 = v_1 + at$	$\omega_2 = \omega_1 + \alpha t$
$s = \frac{1}{2}(v_1 + v_2)t$	$\theta = \frac{1}{2}(\omega_1 + \omega_2)t$
$v_2{}^2 - v_1{}^2 = 2as$	$\omega_2{}^2 - \omega_1{}^2 = 2\alpha\theta$
$s = v_1 t + \frac{1}{2}at^2$	$\theta = \omega_1 t + \frac{1}{2}\alpha t^2$

Example

A flywheel was uniformly accelerated from rest. After two minutes the speed was 30 rev/min. Determine the number of revolutions made by the flywheel in attaining a speed of 120 rev/min.

Solution

$$30 \text{ rev/min} = \frac{2\pi}{60} \times 30 = \pi \text{ rad/s}$$

$$120 \text{ rev/min} = 4\pi \text{ rad/s}$$

First determine the angular acceleration from equation 2.13

$$\omega_2 = \omega_1 + \alpha t \qquad \omega_1 = 0$$

$$\pi = 0 + 120\alpha \qquad \omega_2 = \pi \text{ rad/s}$$

$$\alpha = \frac{\pi}{120} \text{ rad/s}^2 \qquad t = 2 \times 60 \text{ s}$$

$$\alpha = ?$$

Consider now the initial and final states.

$$\omega_3^2 - \omega_1^2 = 2\alpha\theta \quad \text{from equation 2.15}$$

$$16\pi^2 - 0 = \frac{2\pi}{120} \times \theta \qquad \omega_3 = 4\pi \text{ rad/s}, \quad \omega_1 = 0$$

or, $$\theta = \left(\frac{16\pi^2 \times 120}{2\pi}\right) \qquad \alpha = \frac{\pi}{120} \text{ rad/s}^2$$

$$= 960\pi \text{ rad} \qquad \theta = ?$$

so that $$n = \frac{960\pi}{2\pi} \text{ rev} \quad (\text{because } 1 \text{ rev} = 2\pi \text{ rad})$$

and $$n = 480 \text{ rev}$$

Example

A motor car is travelling at 60 km/h when the brakes are applied and the car is brought to rest with uniform retardation in 40 m.

If the outside diameter of the wheels is 50 cm calculate:

(*a*) the linear velocity before the brakes were applied
(*b*) the corresponding angular velocity of the wheels
(*c*) the linear retardation of the car, and
(*d*) the corresponding angular retardation of the wheels

Solution

(*a*) $$60 \text{ km/h} = \frac{60{,}000}{60 \times 60} = 16\tfrac{2}{3} \text{ m/s}$$

(*b*) $$v_1 = r\omega_1 \quad \text{so} \quad \omega_1 = \frac{v_1}{r} \qquad v_1 = \frac{100}{6} \text{ m/s}$$

$$\omega_1 = \frac{100}{6} \times \frac{4}{1} \qquad r = \frac{1}{4} \text{ m}$$

$$= \frac{400}{6} = 66{\cdot}6 \text{ rad/s}$$

(c) $v_2^2 - v_1^2 = 2as$ $\qquad v_2 = 0$

$$0 - \frac{100 \times 100}{6 \times 6} = 2a \times 40 \qquad s = 40 \text{ m}$$

$$a = \frac{-10^4}{36 \times 80} = \frac{-125}{36} = -3{\cdot}47 \text{ m/s}^2$$

So that

Linear retardation $= 3{\cdot}47$ m/s²

(d) $\alpha = \dfrac{a}{r}$

$$= \frac{-125}{36} \times \frac{4}{1} \qquad \left(r = \frac{1}{4}\text{ m}\right)$$

$= 13{\cdot}88$ rad/s² Angular retardation since product is negative

Example

Derive the motion equation $s = ut + \frac{1}{2}at^2$. A vehicle is brought to rest from a speed of 45 mile/h in 60 ft with uniform retardation. The outside diameter of the wheels is 21 in. Find (*a*) the time taken to bring the vehicle to rest, (*b*) the average angular velocity of the wheels, and (*c*) the angular retardation of the wheels. (U.L.C.I.)

Solution

For derivation of the equation, see expression 2.4 of this chapter. Also,

$$v = u + at \quad \text{whence} \quad a = \frac{v}{t} \qquad v = 0$$

$$s = 60 \text{ ft}$$

$$u = 66 \text{ ft/s}$$

(*a*) *a* is a retardation so

$$a = \frac{-u}{t} = \frac{-66}{t}$$

substituting,

$$s - ut + \tfrac{1}{2}at^2$$

$$60 = 66t + \frac{1}{2} \cdot \left(\frac{-66}{t}\right) t^2$$

whence $\quad t = \dfrac{60}{33} = 1{\cdot}82$ s

(*b*) from expression 2.10 above,

$$\omega = \frac{u}{r} \qquad r = \frac{21}{24}\text{ ft}$$

$$\omega_1 = \frac{66 \times 24}{21}\text{ rad/s}$$

$$= 75{\cdot}4\text{ rad/s}$$

Average angular velocity

$$= \left(\frac{\omega_1 + \omega_2}{2}\right) \qquad \omega_2 = 0$$

$$= \left(\frac{75{\cdot}4 + 0}{2}\right)$$

$$= 37{\cdot}7\text{ rad/s}$$

(*c*) From expression 2.12,

$$\alpha = \frac{a}{r} \qquad \text{from } (a)\ a = \frac{-66}{t}$$

$$a = \frac{-66 \times 33}{60}$$

$$= -36{\cdot}3\text{ ft/s}$$

$$\text{and } \alpha = \frac{-36{\cdot}3 \times 24}{21}$$

$$= -41{\cdot}5\text{ rad/s}^2$$

$\therefore$ Angular retardation $= 41{\cdot}5$ rad/s^2

Relative velocity

All motion is relative, and the motion of any particular body can be considered with reference to some other body, which in calculations on relative velocity, can be regarded as fixed. Usually we regard the earth as fixed, and then express the motions of other bodies in terms of their velocities relative to the earth.

If two or more bodies are moving on independent courses, it is sometimes necessary to determine the motion of one relative to the other. The velocity of one body relative, or referred to another is the velocity which the body would appear to have if viewed by an observer on the other body. This relative velocity will be unaltered if we give each body the *same additional velocity*.

If the motions of the two bodies are uniform they can be represented by vectors. Their relative velocity can then be obtained by drawing a vector diagram.

Suppose in figure 2.6, that vectors *a* and *b* represent the motions of two bodies A and B, and that we require to know the velocity of A relative to B.

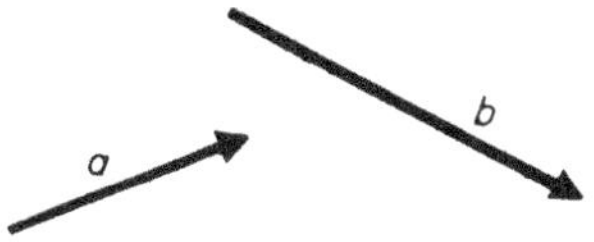

Figure 2.6

The procedure for obtaining the relative velocity is as follows:

1. Reverse the vector *b*.
2. Add this reversed vector to both vectors *a* and *b*.
3. Complete the vector diagram for both vectors *a* and —*b*. Figure 2.7 shows this operation.

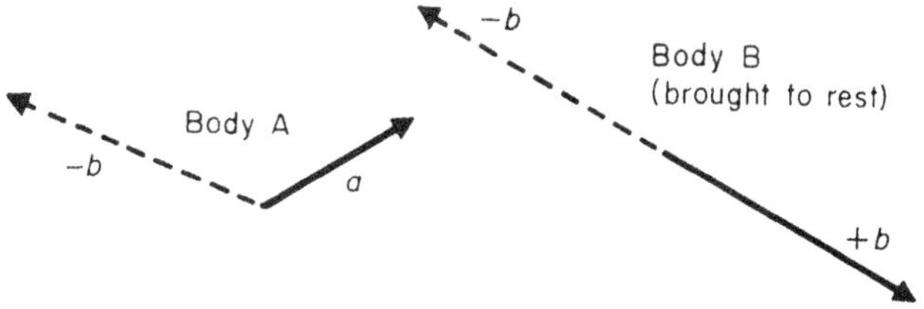

Figure 2.7

The application of the velocity represented by the reversed vector to each of the bodies A and B would not result in any change of their relative motion, and the relative velocities of the bodies would remain unchanged. The effect of adding the vector —*b* to the vector *b*, however, is to reduce this vector to zero: i.e., the body B will theoretically be at rest. In these circumstances the vector sum of *a* and —*b* will give the true velocity of A relative to B. This vector can be obtained by drawing the diagonal of the parallelogram shown in figure 2.8. The diagonal

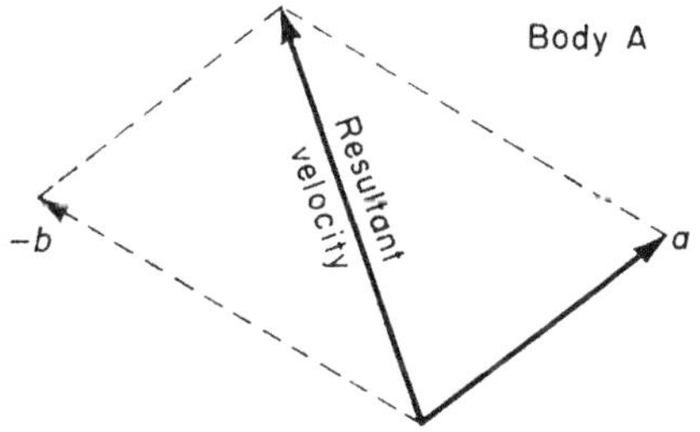

Figure 2.8

represents, in magnitude and direction, the resultant of velocity a and $-b$ in the direction shown.

The velocity of A relative to B can, of course, be found by drawing vectors a and b from a common point, and completing the triangle, the third side representing the relative velocity. Thus in figure 2.9 PRST is the parallelogram obtained from vectors a and $-b$, as shown in figure 2.8, and PS represents the velocity of A relative to B.

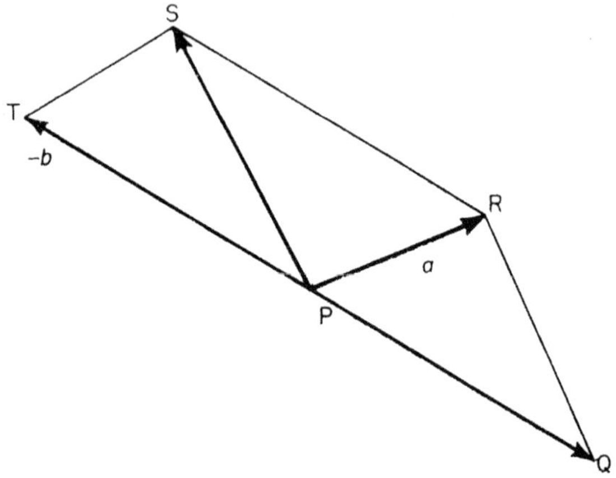

Figure 2.9

If now from P we draw PQ to represent the vector $+b$, and join QR, then QR is equal in magnitude and parallel to PS. Therefore QR also represents the velocity of A relative to B.

The vector diagram

A diagram such as PQR is known as a vector diagram. The interpretation of such a diagram is greatly helped if two simple rules are observed:

1. Any body, which is the reference body for a system, will be given a particular letter—usually O. All other reference points which do not move relative to that body are given the same letter.

2. The motion of one body relative to another is represented by a vector, the ends of which are defined by small letters, corresponding to the capital letters used to denote the bodies.

Thus the motion of body A relative to the point on the stationary body O is represented by the vector oa (figure 2.10). Similarly, the motion of B relative to any point on the stationary body O is represented by the vector ob.

The motion of A relative to the body B is represented by the vector ba: and the motion of B relative to A is represented by vector ab. Consequently, arrows are omitted from such a vector diagram.

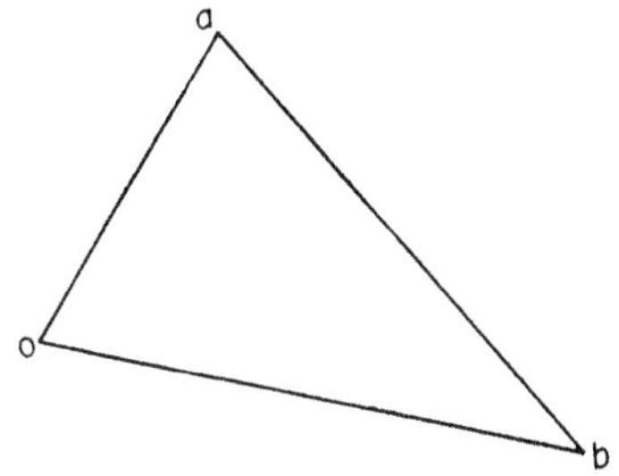

Figure 2.10

Example

A railway carriage rounds a bend in the track, its velocity being 30 ft/s at an angle of 60° West of North. An animal in a field beside the track runs away from the carriage in a due-Northerly direction at a speed of 15 ft/s. What would be the apparent velocity of the animal to an observer in the carriage?

Solution

Let O be the earth, A the animal, and C the carriage. The passenger will have the same velocity as the carriage.

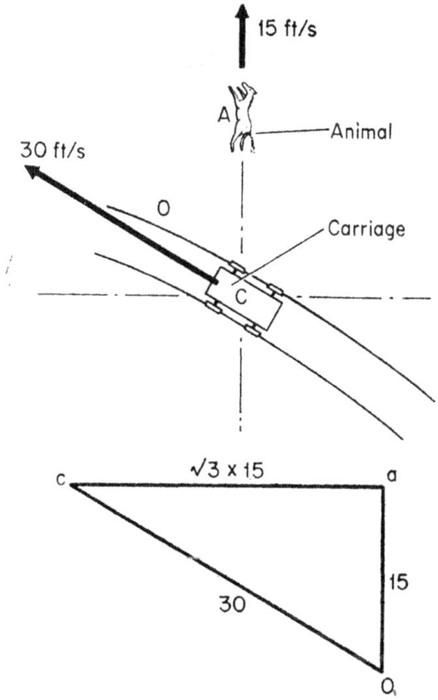

Figure 2.11

Then the velocity of the carriage relative to the earth will be *oc* on the vector diagram. The velocity of the animal relative to the earth will be *oa*, and the velocity of the animal relative to the carriage *ca*. Figure 2.11 shows these relationships.

It can be seen that the diagram is a 60° triangle, with the enclosing sides of the 60° angle having the ratio 1:2. This means that the triangle must be right-angled, and that angle *cao* is the right angle.

From the diagram it can be seen that the direction of the velocity of the animal relative to the passenger in the carriage is due East. The velocity of the animal relative to the passenger in the carriage is $\sqrt{3} \times 15$, or 25·98 ft/s.

Interception. If one body is required to intercept another, then the final position must be that both bodies will be at the same place at the same time. Further, because the relative velocity is unaltered by giving to each body the same additional velocity, we can cause one body to be stopped by applying an equal and opposite velocity to its motion, and add this same velocity vectorially to that of the other body, finally determining the resultant of the two velocity vectors.

Because the first body has been stopped, the second body must proceed from its present position direct to the first body to intercept it. That is, the direction of the relative velocity must be along the line joining the initial positions of the two bodies. Clearly, if the velocities of the two bodies are known, the direction of the path of the intercepting body can be found. A graphical solution is to be preferred.

Example

A rescue launch capable of a speed of 30 knots sets out to intercept a ship 200 miles away to the North, which is steering due East at 20 knots.

(*a*) In what direction should the launch travel?
(*b*) What will be the relative velocity of ship and launch?
(*c*) How long will the interception take?

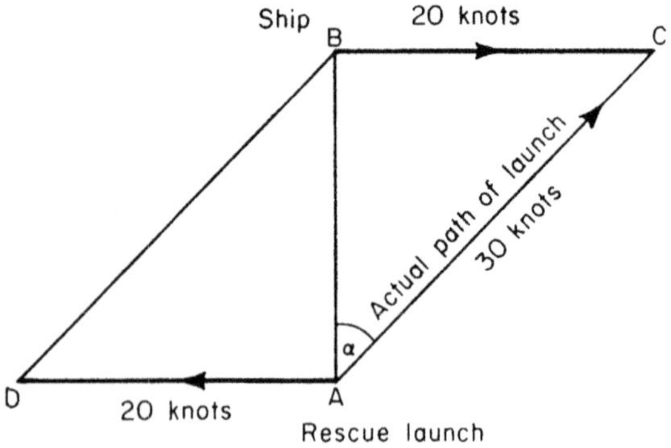

Figure 2.12

Solution

Let A be the position of the rescue launch and B the position of the ship, 200 miles North of A. For interception to take place the relative velocity of A to B must be along AB. Thus if we stop the ship B by applying an equal and opposite velocity of 20 knots to it, we must apply a similar velocity vector to A, along AD. This velocity, combined with the rescue launch velocity of 30 knots along AC will produce a resultant velocity along AB. The angle α, which the actual path of the rescue launch makes with due North is as yet unknown.

Draw BC perpendicular to AB, and equal to 2 inches to represent a speed of 20 knots. From C mark off a length CA equal to 3 inches to cut AB at A, and to represent 30 knots (figure 2.12).

Then from figure 2.12, by measurement,

(*a*) $$\alpha = 42\tfrac{1}{2}°$$

i.e. the launch must travel along a path $42\tfrac{1}{2}°$ East of North.

(*b*) $$AB = 2{\cdot}24 \text{ in.}$$

to the same scale as BC represents 20 knots, the relative velocity of ship to launch is 22·4 knots.

(*c*) $$22{\cdot}4 \text{ knots} = 22{\cdot}4 \times 1{\cdot}15 \text{ mile/h}$$

Therefore, if AB is 200 miles, the time taken by the launch to intercept the ship is the same as the time taken from A to B at 22·4 knots, i.e.,

$$\text{Time taken to intercept} = \frac{200}{22{\cdot}4 \times 1{\cdot}15} = 7{\cdot}8 \text{ h}$$

$$= 7 \text{ h } 48 \text{ min}$$

Summary

Displacement and velocity are vector quantities. They can be represented by straight lines. Distance and speed are different from displacement and velocity in that they need not relate to straight line motion, and are therefore not vector quantities.

For displacement s in time t,

$$\text{Average velocity } v = \frac{s}{t}$$

Speed conversions

$$1 \text{ mile/h} \equiv 1{\cdot}46 \text{ ft/s}, \quad 60 \text{ mile/h} \equiv 88 \text{ ft/s}$$

and $$1 \text{ knot} \equiv 1{\cdot}15 \text{ mile/h} = 1{\cdot}69 \text{ ft/s}$$

The equations of motion

When v_1 = Initial velocity, and v_2 = Final velocity, a = Acceleration, t = Time of motion, and s = Distance travelled, then

$$v_2 = v_1 + at$$

$$s = \tfrac{1}{2}(v_1 + v_2)t$$

$$v_2^2 - v_1^2 = 2as$$

$$s = v_1 t + \tfrac{1}{2}at^2$$

Speed-time graph

The inclination, or slope at a point on a speed-time graph represents the acceleration at that time.

The area under the speed-time graph is equal to the distance travelled.

Vertical motion

In this motion, free-falling bodies experience an acceleration of g, the acceleration due to the earth's gravity, which in London is 32·2 ft/s^2, or 9·81 m/s^2. It is possible to assign to the elements of the motion a sign convention, such that directions upward are positive, and those downward negative. These conventions, applied in the formula above, will allow successful solution of problems on vertical motion.

Angular measure

The radian and the degree are measures of angular displacement. A full circle includes 2π radians or 360°. One radian is therefore equal to 57·3°.

Angular motion

When angular displacement is denoted by θ, angular velocity by ω, r is the radius of the path motion, and angular acceleration is denoted by α, then,

$$v = \omega r, \quad \text{or} \quad \omega = \frac{v}{r} \text{ rad/s}$$

also,

$$a = \alpha r, \quad \text{or} \quad \alpha = \frac{a}{r} \text{ rad/s}^2$$

$$\omega_2 = \omega_1 + \alpha t$$

$$\theta = \frac{\omega_1 + \omega_2}{2} \,.\, t$$

$$\omega_2^2 - \omega_1^2 = 2\alpha\theta$$

$$\theta = \omega_1 t + \tfrac{1}{2}\alpha t^2$$

Relative velocity

All motion is relative. The velocity of one body relative or referred to, another, is the velocity which the body would appear to have if viewed by an observer on the reference body.

To find the velocity of a body A relative to a body B, the velocity of body B is reversed and added vectorially to the vector representing the velocity of body A. The vector sum $v_a + (-v_b)$ is the velocity of body A relative to body B.

Exercises

Take

$$g = 32 \text{ ft/s}^2, \text{ or } 9{\cdot}8 \text{ m/s}^2$$
$$60 \text{ mile/h} = 88 \text{ ft/s}$$
$$100 \text{ knots} = 169 \text{ ft/s}$$

1. What is the acceleration of an aircraft, which, before take-off, increases its speed uniformly from rest to 50 m/s in 30 s?

2. The acceleration of a machine is given as 250 yards per minute each minute. Express this in ft/s². What time will elapse while the machine increases its speed by 25 mile/h?

3. An aircraft in a shallow dive increases its speed from 180 to 350 knots in 4 minutes. Calculate (*a*) the value of the acceleration, assuming it to be constant during the dive, and (*b*) the distance travelled.

4. A car passes a certain point at a speed of 48 km/h, and a second point 2 km away at a speed of 112 km/h. Its acceleration over the measured distance is constant. Calculate (*a*) the average speed of the car, (*b*) the time taken from the first to the second point, and (*c*) the magnitude of the acceleration.

5. A car travelling at 70 mile/h is brought to rest with constant retardation in a distance of 120 yards. Calculate (*a*) the retardation in ft/s², and (*b*) the time taken.

6. A lift rises 40 m in 10 s. During the first quarter of the distance the lift is uniformly accelerated; during the last quarter it is uniformly retarded. Given that its speed remains constant during the middle portion of its ascent, calculate the value of its maximum speed.

7. A stone is dropped from the top of a tower 120 ft high. Calculate (*a*) how far it will have dropped in $2\frac{1}{2}$ seconds, (*b*) how long it will take the stone to reach the base of the tower, and (*c*) the velocity with which the stone strikes the base.

8. A ball is projected vertically upwards with a velocity of 25 m/s. Determine (*a*) the maximum height reached, (*b*) the time taken to reach a height of 30 m, and (*c*) the total time from projection to striking the ground on the return journey.

9. A projectile is fired vertically upwards with a velocity of 900 ft/s from the edge of a cliff 250 ft above the sea. Determine (*a*) the maximum height reached by the projectile, and (*b*) its velocity when it strikes the water, neglecting air resistance.

10. A jet of water issues horizontally from an orifice in the side of a

tank. It is observed to pass a point 20 cm vertically below the centre of the orifice, and 50 cm horizontally from the edge of the tank. Calculate the horizontal velocity of the jet. Take g as 9·8 m/s^2.

11. An electric motor rotating at 3200 rev/min is brought to rest in 15 seconds with a uniform retardation. Calculate: (*a*) the value of the retardation, (*b*) to the nearest whole number, the revolutions made by the motor during the retardation period.

12. A flywheel is accelerated uniformly from rest. After $1\frac{1}{4}$ minutes it is rotating at 170 rev/min. Calculate: (*a*) its angular acceleration (*b*) to the nearest whole number, the revolutions made by the flywheel from rest to a speed of 500 rev/min (*c*) the speed of a point on its periphery when rotating at 500 rev/min, given that the diameter of the flywheel is 1 m.

13. An aircraft, flying due West at 200 knots, is observed 150 miles due East of a target which it has set out to attack. Ten minutes later, a second aircraft, flying at 230 knots, leaves its base 50 miles South-West of the target to engage the attacker.

(*a*) In what direction must the second aircraft fly in order to intercept the attacker?

(*b*) At what distance from the target may contact be expected?

14. A ship is moving North East at 30 km/h, and a second ship appears to an observer on the first to be moving due East at 15 km/h. Determine the actual direction and magnitude of the velocity of the second ship.

15. To an observer on a ship travelling due West at 40 km/h, another ship 4 km due South appears to be travelling North-East at 24 km/h. Determine the magnitude and direction of the velocity of the second ship, and the distance apart of the two ships when nearest to each other.

(E.M.E.U.)

16. What is meant by "relative velocity"?

A vehicle travelling at 35 mile/h on a horizontal track has wheels 26 in. diameter. Assuming the wheels are not slipping, determine:

(*a*) The angular velocity of the wheels.

(*b*) The velocities, in magnitude and direction, of the four points on the rim of one of the wheels which at a given instant are at ends of vertical and horizontal diameters. (E.M.E.U.)

3

Newton's laws of motion

Most of us, at some time or other, have found it necessary to exert some human effort in order to push a motor car, if only for a short distance, along a level road, and experienced the difficulty in starting the movement. Some of us possibly remember how very much more difficult it was to move the old coach-built saloon than the modern mini-car.

This resistance to motion, or even to a change in motion, is possessed by all bodies, but with varying degrees, and is known as INERTIA. The amount of inertia which a body possesses depends on its size and also on what is usually termed "the amount of matter" in the body.

Weight (symbol W)
The weight of a body is the force exerted on it by the earth's gravitational field, and because the force of gravity varies inversely as the square of the distance from the centre of the earth, its value is not constant, even on the surface of the earth.

We need not, however, concern ourselves with the small variations unless we are dealing with inter-space problems. In problems dealing with the weights of bodies on this planet we can assume that these weights are constant.

Weight, being a force, should be expressed in force units. In the past, confusion has been caused by the use of the words "pounds" and "kilogrammes" to denote both force and mass. The notation of B.S. 1991 tried to remedy this by applying the letter "f" to the unit abbreviation to denote forces. This is now being replaced by the SI system, but students encountering the abbreviations lbf and kgf should understand that they mean pounds force and kilogrammes force respectively.

Absolute units of force

The absolute unit of force is that force which would accelerate unit mass at unit rate; e.g. such a force might accelerate one pound mass at one foot per second every second. Unit gravitational forces (in which we measure the weights of bodies) will accelerate unit mass bodies at a rate of g, the acceleration due to the earth's gravity. *The magnitude of a unit of quantity of gravitational force is therefore "g" times the magnitude of the corresponding absolute force unit.*

The force unit of the FPS system is the *poundal*, a unit not frequently used by engineers. But the physicist does find the *dyne*, the force quantity of the CGS system, which will accelerate one gramme mass at one centimetre per second per second, very useful. But the SI system unit of force is the absolute force unit of the MKS system, the *newton*. This is the force which will accelerate one kilogramme mass at one metre per second per second. It is now the standard unit for engineers.

Mass (symbol m)

Mass is a physical property sometimes defined as the quantity of matter or the number of molecules in a body, or as a measure of its inertia. None of these definitions is completely satisfactory, because we cannot visualise the means of measurement. Unlike the weight however, the mass remains constant wherever the body happens to be, and can only be changed by adding to, or taking away from, the quantity of matter in the body.

Strictly speaking, the consciousness of mass derives from weight, and since weight is a force due to gravity, and has in the past taken its units from this phenomenon, the engineer has tended to express mass in a way which will lead to the derivation of results of calculations in a gravitational form. He therefore invented a mass unit which he called the "slug," the use of which gave answers to problems in conventional gravitational units. The adoption of the metric SI system of MKS units is, however, likely to lead to the erosion of this practice.

The weight of a body was expressed in pounds or kilogrammes, and the abbreviations lbf and kgf of B.S. 1991 distinguished the units as forces. The mass was also expressed in pounds or kilogrammes, but the standard abbreviations for these were lb and kg. This book is published at a time of transition of unit systems, so both the slug system of mass, where mass is expressed in terms of weight, and the absolute system of mass, will be explained. The kilogramme (kg) is the SI mass unit.

On the slug system,

$$\text{Mass} = \frac{\text{Weight}}{g} = \frac{W}{g} \text{ (slugs)}$$

On the absolute system,

$$\text{Mass} = m \text{ (lb or kg)}$$

The calculations below will use gravitational systems for English units, and absolute systems for SI units. It should be noted, therefore, that the symbols $\frac{W}{g}$ and m are used interchangeably for mass, according to whether the example is to be worked in gravitational (English) units or absolute (SI) units.

Momentum

If a body is already in motion, it must be moving with a finite velocity. The product of the mass and this velocity is called the *momentum* of the body.

$$\text{Momentum} = \text{Mass} \times \text{Velocity}$$

$$= m \times v \quad \text{or} \quad \frac{W}{g} \times v \qquad 3.1$$

Units of momentum

The value of the gravitational expression of mass was partly that its constituent units could be cancelled with others in expressions to give the units of the answer, or as checks on calculations. This is not possible with absolute units. On the slug system,

$$\begin{aligned} \text{Momentum} &= \text{Mass (slugs)} \times \text{Velocity (feet/second)} \\ &= \text{Momentum (slug-feet/second)} \\ &= \text{Momentum (lbf ft}^{-1}\text{ s}^2 \times \text{ft s}^{-1}) \\ &= \text{Momentum (lbf s)} \end{aligned}$$

On the absolute system momentum is of course in kg m s^{-1} units.

Example

A forging hammer of mass 1,300 kg falls freely from a height of 5 m on to a steel ingot. Calculate the momentum of the hammer at the moment of impact. Take $g = 9{\cdot}81$ m/s^2.

Solution

We must first find the velocity of the hammer when it strikes the ingot. Let v_2 be this velocity. The equation of motion is

$$\begin{aligned} v_2^2 - v_1^2 &= 2as & v_1 &= 0 \\ v_2^2 - 0 &= 2 \times 9{\cdot}81 \times 5 & v_2 &= ? \\ v_2^2 &= 98{\cdot}1 & s &= 5 \text{ m} \\ v_2 &= 9{\cdot}9 \text{ m/s} & a &= 9{\cdot}81 \text{ m/s} \end{aligned}$$

(*Note:* All the quantities are measured in the same direction, so all may be regarded as positive whatever convention is used.)

Momentum of hammer at impact

$$= \text{Mass} \times \text{Velocity}$$
$$= m \times v_2 \text{ (abs)}$$
$$= 1{,}300 \times 9{\cdot}9$$
$$= 12{,}870 \text{ kg m s}^{-1}$$

Example

A motor-car of mass 1,500 lb is travelling along a road at a speed of 20 mile/h. After a few seconds its speed has increased to 40 mile/h. Calculate the change in its momentum.

Solution

$$60 \text{ mile/h} = 88 \text{ ft/s}$$
$$20 \text{ mile/h} = \frac{88}{3} \text{ ft/s} = v_1$$
$$40 \text{ mile/h} = \frac{176}{3} \text{ ft/s} = v_2$$
$$\text{Momentum} = \text{Mass} \times \text{Velocity}$$
$$\text{Change in momentum} = \text{Mass} \times \text{Change in velocity}$$
$$= \text{Mass} \times (v_2 - v_1)$$
$$= \text{Mass} \times \left(\frac{176}{3} - \frac{88}{3}\right)$$
$$= \frac{1{,}500}{32} \times \frac{88}{3} \text{ (grav)}$$
$$= 1{,}375 \text{ lbf s}$$

Conservation of linear momentum

When two bodies, each possessing momentum, are moving in the same straight line, and collide, the total momentum of the system remains constant before and after impact. This rule is called *the principle of conservation of linear momentum.*

If masses m_1, m_2, moving with velocities u_1, u_2, collide, and move subsequently in the same straight line with velocities v_1, v_2, then

$$m_1u_1 + m_2u_2 = m_1v_1 + m_2v_2$$

The principle can be proved as follows, using Newton's 3rd law of motion (p. 35).

Consider the change of momentum of each body. The impulse I_1 (see p. 40), which m_1 receives during the impact, is equal and opposite to the impulse I_2, received by m_2, because the time of contact and the force between them is the same for each. Then,

$$I_1 = m_1v_1 - m_1u_1$$

and
$$I_2 = m_2v_2 - m_2u_2$$

adding these expressions,

$$m_1u_1 + m_2u_2 = m_1v_1 + m_2v_2 \qquad 3.2$$

This result would still apply, with appropriate sign changes, if the directions of motion of the masses in the same straight line were opposite to each other.

Example

A ball-bearing of mass 100 g, rolling down a groove at a velocity of 500 cm/s, squarely strikes a stationary ball-bearing of mass 50 g. The velocity of the first ball-bearing after impact is 200 cm/s in the direction of its motion before impact. What is the velocity of the second?

Solution

Using the notation of 3.2 above,

$$m_1u_1 + m_2u_2 = m_1v_1 + m_2v_2 \qquad m_1 = 0{\cdot}1\text{ kg}$$
$$(0{\cdot}1\text{ kg} \times 5\text{ m/s}) + (0{\cdot}05\text{ kg} \times 0) \qquad u_1 = 5\text{ m/s}$$
$$= (0{\cdot}1\text{ kg} \times 2\text{ m/s}) + (0{\cdot}05\text{ kg} \times v_2) \qquad m_2 = 0{\cdot}05\text{ kg}$$
$$0{\cdot}5 + 0 = 0{\cdot}2 + 0{\cdot}05v_2 \qquad u_2 = 0$$
$$0{\cdot}05v_2 = 0{\cdot}3 \qquad v_1 = 2\text{ m/s}$$
and
$$v_2 = 6\text{ m/s} \qquad v_2 = ?$$

Newton's laws of motion

Almost three hundred years ago Sir Isaac Newton stated certain relationships which have since become known as Newton's laws of motion. They are expressed as follows:

1. 'A body remains in a state of rest, or uniform motion in a straight line unless it is acted upon by an external force.'

2. 'The rate of change of momentum of a body is equal to the applied force, and takes place in the direction of the applied force.'

This law can also be stated in the form of an expression which will be familiar to most of us, namely

$$\text{Applied force} = \text{Mass} \times \text{Acceleration}$$

or,
$$P = \frac{W}{g} \times a \quad \text{or} \quad = m \times a \qquad 3.3$$

3. 'To every action there is an equal and opposite reaction.'

The first law

A body is either at rest or in motion. The first law states that if a body is initially at rest, it will remain in this state for all time unless some external force acts upon it and causes it to move. Thus, suppose a motor-car is stationary on a level road and its engine is shut off: unless the car is pushed or pulled by some external force, it will remain stationary in the same position.

If a body is in motion at any instant—that is, moving with a given velocity—then, according to the first law of motion, it will continue to move in a fixed direction with the same velocity unless some external force acts upon it and causes it to change its velocity either in direction or magnitude.

It is not easy to select an example of the first law applied to a body in motion, because any body moving on this earth or in the atmosphere surrounding it, is resisted by friction forces and also affected by the earth's gravitation. It is possible, however, to conceive of a body moving in outer space well beyond the Universe as we know it, and relatively free of the gravitational attractions of its members. There might then be only negligible external forces acting on it, so that it might continue its motion with constant velocity in a fixed direction.

This property of a body to remain in a given state unless acted upon by an external force, is known as 'inertia'. Because of its inertia a body will be sluggish in changing its state, and will resist any change. Newton's first law is often referred to as the 'Law of Inertia'.

If the mass and the velocity both remain constant, we have, from equation 3.1,

$$\text{Momentum} = m \times v = \text{Constant}$$

The first law may, therefore be expressed in terms of a body moving with constant momentum.

The second law

Suppose that a heavy block of steel of weight W is resting on a horizontal surface, and a force (figure 3.1) is applied to it, tending to push it along the table.

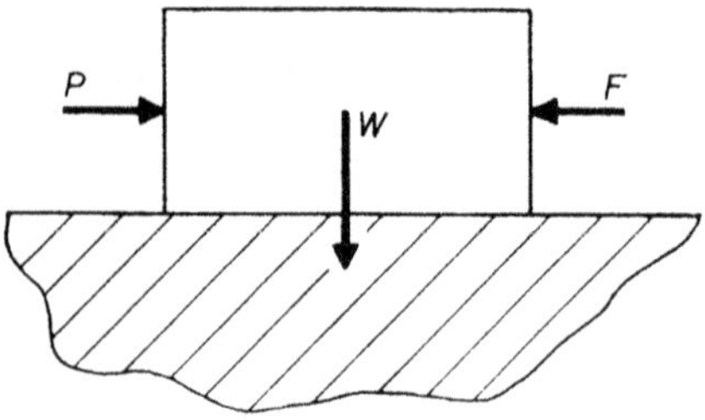

Figure 3.1

If there is no resistance to the motion of the block, that is, it can slide freely over a smooth surface, the block will accelerate in the direction of the applied force P in accordance with equation 3.3.

Suppose, however, that an opposing force F in the form of a resistance is met. If the magnitude of this resistance is less than P, the body will continue to accelerate in the direction of P, but the magnitude of the acceleration will be less.

If the resistance is equal in magnitude to P, the body will remain stationary, or, if given a motion, will continue with uniform velocity in the direction of the applied force.

The force which produces the acceleration of the body is, therefore, not equal to the applied force if the motion is resisted, but rather to the net force, or effective force, where

$$\text{Effective force} = (P - F)$$

Hence, in practice, a more general form of Newton's second law is:

$$\begin{aligned}\text{Net, or Effective Force} &= \text{Force producing acceleration,}\\ &= m \times a\end{aligned}$$

At first sight, the two ways of stating Newton's second law:

$$\text{Rate of change of momentum} = \text{Applied force}$$

and

$$\text{Applied force} = \text{Mass} \times \text{Acceleration}$$

are not the same. Let us examine more closely what is meant by the rate of change of momentum.

$$\text{Rate of change of momentum} = \frac{\text{Change of momentum}}{\text{Time taken for the change}}$$

If suffixes (1) and (2) denote initial and final states, then,

$$\begin{aligned}\text{Rate of change of momentum} &= \frac{(m\,v_2 - m\,v_1)}{t}\\ &= m\left(\frac{v_2 - v_1}{t}\right)\\ &= m \times a\\ &= P \text{ (the applied force)} \qquad 3.4\end{aligned}$$

Note that the acceleration is the rate of change of velocity, and takes place in a direction which may be obtained by drawing a velocity vector triangle. Momentum is the product of a constant and the velocity, so that it also can be represented by a vector. Its vector triangle will be similar in all respects to the velocity vector triangle, and so change of momentum will take place along the same line as the change in velocity.

Example

The table of a planing machine is 250 kg mass, and accelerates from rest to a speed of 15 m/min in $1\frac{1}{2}$ s. Calculate the average force required to accelerate it during this portion of its travel. Take $g = 9{\cdot}81$ m/s^2.

Solution

$$\text{Average force} = \frac{\text{Change of momentum}}{\text{time}}$$

$$\text{Change of momentum} = m\,(v_2 - v_1) \qquad v_1 = 0$$

$$= 250\left(\frac{15}{60} - 0\right) \qquad v_2 = \frac{15}{60}\text{ m/s}$$

$$= \frac{250}{4} \qquad g = 9{\cdot}81\text{ m/s}^2$$

$$= 62{\cdot}5 \qquad m = 250\text{ kg}$$

$$\text{Average force} = 62{\cdot}5 \times \tfrac{2}{3} \qquad t = 1{\cdot}5\text{ s}$$

$$= 41{\cdot}67\text{ N}$$

Example

A lift weighs 2,400 lbf when fully loaded, and the frictional resistance to its motion is 110 lbf. Determine the tension in the rope:

(*a*) When the lift is being accelerated upwards at 4 ft/s^2.
(*b*) When the lift is being raised at constant velocity.
(*c*) When the lift is being lowered with an acceleration of 2 ft/s^2.

Solution

Let P = Force, or tension in the rope.
W = Weight of lift acting vertically downwards.
F = Frictional resistance, acting in the direction opposite to the motion.

(*a*) Net, or effective force upwards $= P - (W + F)$

This is the force producing an acceleration upwards.

From Newton's second law,

$$\text{Effective force} = \text{Mass} \times \text{Acceleration}$$

So that

$$P - (W + F) = \frac{W}{g} \times a$$

$$P - (2{,}400 + 110) = \frac{2{,}400}{32} \times 4 \qquad W = 2{,}400\text{ lbf},\quad F = 110\text{ lbf}$$

$$P - 2{,}510 = 300 \qquad a = 4\text{ ft/s}^2$$

$$P = 2{,}810\text{ lbf} \qquad g = 32\text{ ft/s}^2$$

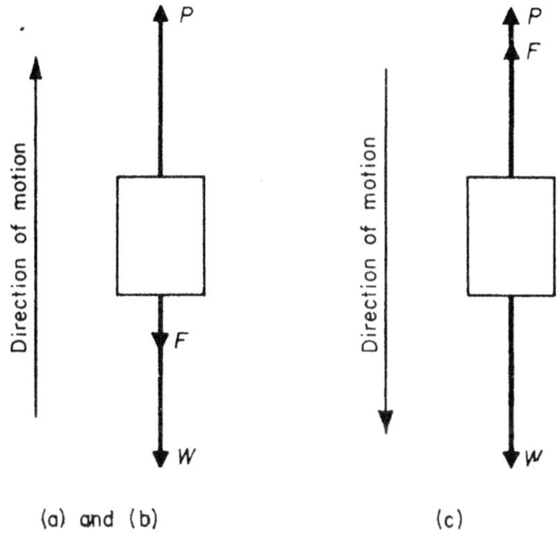

Figure 3.2 Lift motion

(*b*) If the lift is moving upwards with constant velocity, the value of the acceleration is zero. We can still use Newton's second law and say

$$\text{Force producing acceleration} = 0$$
$$P - (W + F) = 0$$
$$P = W + F$$
$$= 2{,}400 + 110$$
$$= 2{,}510 \text{ lbf}$$

(*c*) When the lift is descending, its weight must be greater than the combined effect of the tension in the rope and the frictional resistance. Consequently,

$$\text{Net, or effective force downwards} = W - (P + F)$$

This is the force producing an acceleration downwards, Then,

$$W - (P + F) = \frac{W}{g} \times a$$

$$2{,}400 - P - 110 = \frac{2{,}400}{32} \times 2 \qquad a = 2 \text{ ft/s}^2$$

or,

$$P = 2{,}400 - 110 - 150$$

$$P = 2{,}140 \text{ lbf}$$

Impulse

We have seen that

$$\text{Rate of change of momentum} = m\left(\frac{v_2 - v_1}{t}\right)$$

and that the rate of change of momentum = applied force.

$$\text{Applied force} = \frac{\text{Change of momentum}}{\text{Time taken}}$$

$$\therefore \text{ Applied force} \times \text{Time taken} = \text{Change of momentum}$$

$$P \times t = m\,(v_2 - v_1) \qquad 3.6$$

If a very short interval of time is taken the force can be assumed constant. The expression ($P \times t$) is then known as the "Impulse," and is equal to the change of momentum of the body during the time interval t. An impulsive force means a sudden blow, or a sudden application of the force.

Example
The tup of a drop hammer is travelling at 10 m/s when it strikes. It is brought to rest in 0·2 s. The tup weighs 20 kgf. What is the force of the blow?

Solution

$$\text{Average force} = \text{Mass} \times \frac{\text{Change of velocity}}{\text{Time taken}}$$

$$= 20 \text{ kg} \times \frac{10 \text{ m/s}}{0{\cdot}2 \text{ s}}$$

$$= 1{,}000 \text{ kg m/s}^2$$

$$= 1{,}000 \text{ N}$$

(*Note* that N is the abbreviation for newton.)

Example
A railway truck of mass 20 tonne travelling at 4 m/s collides with a truck of mass 12 tonne(t) travelling in the same direction at 3 m/s. The trucks are coupled during impact.

(*a*) Determine their common velocity and kinetic energy after impact.

(*b*) If the trucks are brought to rest by buffers in a distance of 150 mm determine the average resistance exerted by the buffers on the trucks. (U.L.C.I.)

Solution
From expression 3.2 (p. 34):

$$
\begin{aligned}
m_1u_1 + m_2u_2 &= m_1v_1 + m_2v_2 & m_1 &= 20\ \text{t}\\
(20 \times 4) + (12 \times 3) &= (20 + 12)v & m_2 &= 12\ \text{t}\\
80 + 36 &= 32v & u_1 &= 4\ \text{m/s}\\
v &= \frac{116}{32} & u_2 &= 3\ \text{m/s}\\
&= 3{\cdot}63\ \text{m/s} & v_1 &= v_2 = v,\ \text{say}
\end{aligned}
$$

Kinetic energy (see p. 45)

$$
\begin{aligned}
&= \frac{1}{2} m \times v^2 & m &= 20 + 12 = 32\ \text{t}\\
&= \frac{32 \times 3{\cdot}63^2}{2} & v &= 3{\cdot}63\ \text{m/s}\\
&= 210{\cdot}8\ \text{kJ}
\end{aligned}
$$

Let

Average buffer force $= F$

Buffers absorb all Kinetic Energy over 150 mm distance.

$$
\begin{aligned}
F \times 0{\cdot}15\ \text{m} &= 210{\cdot}8\ \text{kJ}\\
F &= 1405\ \text{kN}
\end{aligned}
$$

Summary

The *weight* of a body is the force of the earth's gravitational attraction on its mass. This is the best index of the amount of matter in the body (its mass). Engineers have used this fact to derive a unit of mass in terms of force and acceleration.

$$\text{Mass} = \frac{\text{Weight}}{g} = \frac{\text{lbf}}{\text{ft/s}^2} = \frac{\text{lbf s}^2}{\text{ft}}$$

These units are called 'slugs' on the FPS system.

If mass is expressed in lb or kg, the derived answer will be in absolute units.

Momentum

This is the product (mass × velocity) of a body.

The *Units of momentum* are derived from this product. If the slug unit for mass is used, then the momentum unit is given by (slugs × ft/s), or slug ft/s. This unit, cancelled down, gives $\frac{\text{lbf}}{\text{ft}} \times \text{s}^2 \times \frac{\text{ft}}{\text{s}}$, or lbf s.

If mass is expressed in lb or kg, momentum derived is lb ft s^{-1} or kg m s^{-1}.

Conservation of linear momentum

When two bodies, each possessing momentum, are moving in the same straight line and collide, the total momentum of the system remains theoretically constant before and after impact. In symbols,

$$m_1u_1 + m_2u_2 = m_1v_1 + m_2v_2$$

Newton's laws of motion

These can be expressed:

First law. A body remains in a state of rest, or uniform motion in a straight line unless acted upon by an external force.

Second law. The rate of change of momentum of the body is equal to the applied force, and takes place in the direction of the applied force.

Third law. To every action there is an equal and opposite reaction.

From the second law we derive the important relationship:

$$\text{Applied force } P = \text{Mass } \frac{W}{g} \times \text{Acceleration } a$$

$$P = \frac{W}{g} \times a$$

Absolute units of force

The absolute unit of force is that force which will impart unit acceleration to unit mass. Thus,

The poundal will accelerate one pound mass at one foot per second per second.

The newton will accelerate one kilogramme mass at one metre per second per second.

A note on units of force

In the FPS system of units, gravitational units have traditionally been more important for purposes of calculation by engineers. But the adoption of the MKS system by electrical engineers has meant that the units of this system, with their international applicability have become more important in engineering design. The newton, which is the absolute unit of force on the SI system, is important.

Exercises

Take

$$g = 32 \text{ ft/s}^2, \text{ or } 9{\cdot}8 \text{ m/s}^2$$

$$1 \text{ (metric) tonne} = 1000 \text{ kg, abbreviated } 1 \text{ t}$$

1. A car of mass 1 tonne is accelerated from rest on a horizontal track. In a distance of 0·5 km it reaches a speed of 110 km/h. Calculate: (*a*) the time taken from rest to reach this speed, (*b*) the acceleration, and (*c*) the force exerted by the car, neglecting air resistance and friction losses.

2. A train is 250 t mass, and the frictional resistance to its motion is 98·1 N/t. Find the value of the force exerted by the engine to maintain constant speed on a level track.

3. A train weighing 200 tonf requires a thrust of 15 tonf to accelerate at 2 ft/s^2 on a horizontal track. Calculate the frictional resistance in lbf/tonf of train weight.

4. A pit cage complete with load is 3 t mass. It is uniformly accelerated upwards by a rope to a speed of 6 m/s in a height of 35 m. Find the tension in the rope.

5. Determine the tension of the rope of a 10 cwt lift when the lift is (*a*) accelerating upwards at 2·5 ft/s^2, (*b*) travelling with uniform speed, and (*c*) accelerating downwards at 2·5 ft/s^2.

6. Whilst travelling along a horizontal road at 24 km/h, a cyclist suddenly stops pedalling, and after 10 s finds that his speed has fallen to 12 km/h. What is the force resisting his forward motion? The mass of the cyclist and bicycle together is 80 kg.

4

Mechanical energy

Energy is usually defined as the capacity for doing work. There are many forms of energy, but in this chapter we shall only be concerned with mechanical energy.

Mechanical energy is the capacity possessed by a body to do mechanical work. We shall, therefore, be thinking of a store of energy, which, when released, can readily be converted into mechanical work. Consequently, energy and work done will be interdependent. We shall say, in fact, that

$$\text{Change of energy} = \text{Work done}$$

and both energy and work done will have the same units.

We have already seen that work involves a force and a movement, and that no work is done until movement takes place, whatever the magnitude of the force. Consequently, when we are thinking of mechanical energy, we shall have in mind a movement, a change of movement, or a prospective movement.

Mechanical energy can exist in two forms. The body may have a capacity for doing work either because it has a height advantage—that is it is initially at some higher level than the new level we are using as reference—or, because it has a velocity.

Potential energy

A body is said to acquire potential energy when it is raised vertically, thus giving it a height advantage.

Work is done on the body by raising it against the force of gravity, and so an extra store of energy is acquired by the body equivalent to the external work done on it. Thus if the body weighs W and is raised h above some agreed datum,

$$\text{Work done} = W \times h$$

Hence,

$$\text{Potential energy acquired by body} = Wh \qquad \textbf{4.1}$$

Kinetic energy

If a body weighing W $\left(\text{i.e. having a mass of } \frac{W}{g}\right)$ is moving with a velocity v in a given direction, it is said to have kinetic energy.

To determine the value of this kinetic energy, let us suppose that the body has been given its velocity by an external force P acting on it over a distance s. Then the work done by the force must be stored in the moving body as kinetic energy. But,

$$\text{Work done by force} = \text{Force} \times \text{Distance moved in direction of force}$$
$$= P \times s$$

and $$\text{Force} = \text{Mass} \times \text{Acceleration}$$

or, $$P = \frac{W}{g} \times a \text{ (gravitational)}$$

so that $$a = \frac{Pg}{W} \left(\text{or } \frac{P}{m} \text{ absolute}\right) \qquad 4.2$$

The velocity of the body after travelling s ft may be obtained from the equation

$$v_2^2 - v_1^2 = 2as \qquad 4.3$$

and $$s = \frac{v_2^2}{2a}, \text{ since } v_1 = 0$$

We can now omit the suffix 2 without losing any clarity.

Substituting the value of a from 4.2, we have

$$s = \frac{v^2}{2} \times \frac{W}{Pg}$$

and

$$\text{Work done by force} = P \times \frac{v^2}{2} \times \frac{W}{Pg}$$
$$= \frac{1}{2}\frac{W}{g} v^2 \text{ (gravitational)}$$

i.e.

$$\text{Kinetic energy acquired by moving body} = \frac{1}{2}\frac{W}{g} v^2 \text{ (gravitational)} \qquad 4.4$$

or, $$= \tfrac{1}{2}mv^2 \text{ (absolute)} \qquad 4.4$$

The units of mechanical energy

We have seen that energy is convertible into work, and therefore we must expect the units of mechanical energy and work to be the same. Substituting FPS units in the expression for kinetic energy, we get

$$\text{Kinetic energy} = \frac{1}{2}\frac{W}{g}\ \text{lbf ft}^{-1}\ \text{s}^2 \times v^2\ \text{ft}^2\ \text{s}^{-2}$$

the units being $\dfrac{\text{lbf} \times \text{s}^2 \times \text{ft}^2}{\text{ft} \times \text{s}^2}$ i.e., ft lbf.

(*Note:* If the mass m had been used in the kinetic energy expression (i.e. K.E. $= \frac{1}{2}mv^2$), then the SI energy unit derived would be the metre-newton, or joule. This unit, abbreviated Nm, is the standard SI work unit.)

The SI work unit. The metre-newton, or joule

The work unit of the MKS system is, of course, the product of unit distance and unit force on that system.

1 metre × 1 newton = 1 metre-newton (abbreviated Nm)

This unit, the metre-newton, is called the *joule*. The joule is of great importance owing to its association with the electrical power unit, the watt, which is a rate of working of one joule per second, and which is now the standard SI power unit.

Example

A car of mass 1,000 kg and travelling at 45 km/h is braked to a standstill over a distance of 40 m. Assuming that the braking force is constant, calculate the road force at each tyre.

Solution

$$45\ \text{km/h} = 45\ \text{km} \times \frac{1{,}000\ \text{m}}{1\ \text{km}} \times \frac{1\ \text{h}}{3{,}600\ \text{s}} = 12\tfrac{1}{2}\ \text{m/s}$$

$$\text{Initial K.E. of car} = \tfrac{1}{2} \times 1{,}000 \times (12\tfrac{1}{2})^2\ \text{joules}$$

$$= 78{,}125\ \text{J}$$

$$\text{Total road force} \times 40\ \text{m} = 78{,}125\ \text{J}$$

$$\text{Road force} = \frac{78{,}125\ \text{J}}{40\ \text{m}} = 1953{\cdot}1\ \text{N}$$

$$\text{Force per tyre} = \frac{1953{\cdot}1}{4} = 488{\cdot}3\ \text{N}$$

Conservation of energy

There is an important empirical law which states that energy cannot be created or destroyed. In other words, the total quantity of energy available at any time must be constant. Energy can, however, change its form.

As a consequence of this law we can state that a body which possesses potential energy because of its height above the ground loses some of this energy when it falls to a lower height. In doing so, however, it acquires velocity, and so we can say

$$\text{Loss of potential energy} = \text{Gain in kinetic energy} + \text{Work done}$$

Example

A weight of $\frac{1}{4}$ tonf, after falling freely from a height of 12 ft, transfers some of its energy to a machine, and in doing so its velocity is decreased to 15 ft/s. Calculate the amount of energy given to the machine.

Solution

$$\begin{aligned}\text{Loss of potential energy by weight} &= Wh\\ &= \frac{2{,}240 \times 12}{4}\ \text{ft lbf}\\ &= 6{,}720\ \text{ft lbf}\end{aligned}$$

Let E be the energy transferred to the machine: then, from the conservation of energy law,

$$\begin{aligned}\text{K.E. of weight after falling 12 ft} &= \text{Energy transferred to machine}\\ &\quad + \text{K.E. of weight at 15 ft/s}\\ 6{,}720 &= E + \left(\frac{1}{2} \times \frac{2{,}240}{4 \times 32} \times 15^2\right)\\ &= E + 1{,}970\\ \text{i.e.} \qquad E &= 6{,}720 - 1{,}970\\ &= 4{,}750\ \text{ft lbf}\end{aligned}$$

Summary

Energy

Energy is capacity for doing work. This chapter has been concerned only with mechanical forms of energy.

Potential energy is that form in which energy is possessed by a body owing to its height. This energy is recoverable through the normal action of gravity on the body, during a fall to a lower position.

$$\text{Potential energy} = \text{Weight} \times \text{Height} = Wh$$

Kinetic energy is possessed by a body due to its motion. A body of weight W, moving with velocity v, will have

$$\text{Kinetic energy} = \frac{1}{2}\frac{W}{g}v^2$$

or, for mass m, $\frac{1}{2}mv^2$.

Conservation of energy

Energy cannot be created or destroyed. Where the fall of a mass causes work to be done we can say,

$$\text{Loss of potential energy} = \text{Gain of kinetic energy} + \text{Work done}$$

The SI unit of work

The SI unit of work (or energy) is the metre-newton, or joule.

Exercises

1. A tup, of mass 100 kg is raised 50 m above the ground. It is then allowed to fall freely on to a pile the top of which is 6 m above the ground. Calculate (*a*) the potential energy of the tup at its highest point, (*b*) its potential and kinetic energies when it strikes the pile.

2. A bomb of mass 200 kg is dropped from an aircraft flying at 80 m. Assuming the air offers no resistance, calculate the velocity with which the bomb strikes the ground, and its kinetic energy on striking. What was its potential energy at the moment of release from the aircraft?

3. A weight of 200 lbf falls freely from rest through a vertical height of 25 ft. Its velocity is then suddenly reduced to 20 ft/s by the loss of some of its energy to a machine. Calculate the energy given to the machine.

4. A motor-car of mass 1 t is travelling at 90 km/h. What is its kinetic energy at this speed? Determine the loss of kinetic energy when its speed has been reduced to 40 km/h.

5. The driver of a motor-car weighing 1 tonf applies his brakes and reduces his speed from 45 mile/h to 10 mile/h in 5 s. Find the loss of kinetic energy of the car.

6. A motor-car of mass 1000 kg is travelling on a level road at 40 km/h. Find (*a*) its kinetic energy at this speed, and (*b*) the force, assumed constant, necessary to raise its speed to 90 km/h in 10 s.

7. A train weighing 250 tonf and travelling at 45 mile/h is brought to rest in a distance of $1\frac{3}{4}$ miles by a constant braking force. What is the value of this force? Determine also the change in the kinetic energy of the train, and the work done in bringing it to rest.

5

Motion in a circular path

We have already shown that, if a body is rotating in one plane with a constant angular velocity ω about a fixed point, its linear velocity is determined by the expression

$$v = r\omega \qquad 5.1$$

Kinetic energy of rotation

Consider a particle of weight W rotating with constant angular velocity ω about a centre O at radius r (figure 5.1).

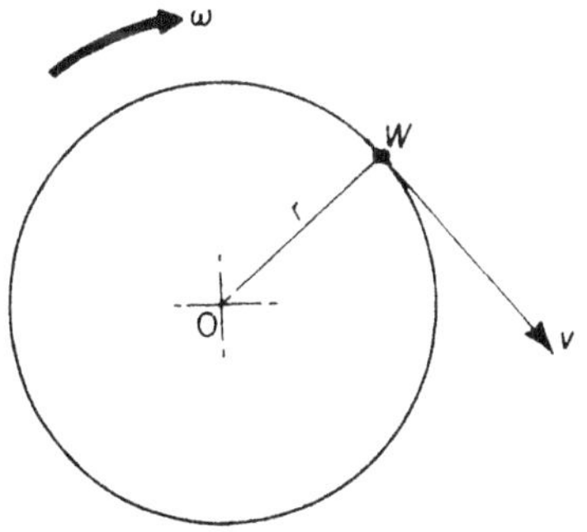

Figure 5.1

Its linear velocity at any instant is given by $v = r\omega$. Its kinetic energy will be given by

$$\begin{aligned} \text{K.E.} &= \frac{1}{2}\frac{W}{g}\; v^2 \\ &= \frac{1}{2}\frac{W}{g}\; r^2\omega^2 \\ &= \frac{1}{2}\,\omega^2\left(\frac{W}{g}\,r^2\right) \end{aligned}$$

Now suppose we have a body composed of a large number of particles of weights W_1, W_2, W_3, etc., acting at radii r_1, r_2, r_3, etc., from the common centre of rotation O. Then the

$$\text{Total kinetic energy} = \frac{1}{2}\,\omega^2\left(\frac{W_1}{g}\,r_1^2 + \frac{W_2}{g}\,r_2^2 + \frac{W_3}{g}\,r_3^2 + \ldots\right)$$

Suppose $\qquad W_1 + W_2 + W_3 + \ldots = W$

where W is the weight of the whole rotating mass. We can consider this mass to be concentrated at some radius k such that its effect on rotation would be equal to the effects of all the small particles rotating separately at a common angular velocity.

Hence,

$$\text{Total kinetic energy of the rotating mass} = \frac{1}{2}\,\omega^2\left(\frac{W}{g}\,k^2\right)$$

The expression $\left(\frac{W}{g}\,k^2\right)$ is known as the *moment of inertia* of the body, and k is known as the *radius of gyration.*

The symbol used for the moment of inertia is I, so that

$$I = \frac{W}{g}\,k^2 = mk^2 \qquad 5.2$$

If W is measured in lbf and k in ft units, the unit of I will be slugs ft^2.

$$\text{K.E. of rotation} = \frac{1}{2}\,\omega^2\left(\frac{W}{g}\,k^2\right)$$
$$= \tfrac{1}{2}I\omega^2 \text{ ft lbf (FPS)} \qquad 5.3$$

(*Note:* If the mass is expressed in absolute units (kg), then

$$\text{K.E.} = \tfrac{1}{2}\omega^2 mk^2$$

This expression, on the SI system, gives an answer in joules.)

The flywheel

We have just seen that the kinetic energy of rotation of a mass can be determined from the expression

$$\text{K.E. of rotation} = \frac{1}{2}\,I\omega^2 = \frac{1}{2}\left(\frac{W}{g}\,k^2\right)\omega^2$$

In other words, a heavy mass rotating at a high speed acts as a great store of energy. This energy can readily be made available for doing work.

A flywheel is a heavy mass of iron, with much of its mass concentrated at the rim. It possesses two important properties. First, being a heavy mass, that is, possessing considerable inertia, it requires much effort to move it or to change its motion, and any change takes place slowly. Second, if it is rotating at high speed, it becomes a reservoir of energy. These two properties are extremely useful when the flywheel is attached to an engine shaft.

Take, for example, a four-stroke engine. Energy is supplied to the piston only once in every four strokes, and even throughout the power stroke the supply of energy varies considerably. A heavy flywheel attached to the output shaft would tend to even out the fluctuations because it would absorb and store up energy while the engine output was more than sufficient to meet the demand, and would give up energy during the period when the engine was idling.

Example

An engine flywheel, with most of its mass concentrated at the rim is 5 t mass. The mean radius of the rim is 2 m. Calculate the energy stored in the flywheel when it is rotating at 150 rev/min.

Solution

$$\omega = \frac{2\pi n}{60} \qquad n = 150 \text{ rev/min}$$

$$= 2 \times \frac{22}{7} \times \frac{150}{60}$$

$$= \frac{110}{7} \text{ rad/s}$$

If we assume that the centroid of a cross-section of the rim lies on the mean radius, then

$$\text{K.E. of rotation} = \frac{1}{2} m k^2\omega^2 \qquad W = 5 \times 1{,}000 \text{ kgf}, \quad m = 5{,}000 \text{ kg}$$

$$= \frac{1}{2} \times 5{,}000 \times 2^2 \times \frac{110^2}{7^2} \qquad k = 2 \text{ m}$$

$$= 2{\cdot}46 \times 10^6 \text{ J}$$

Centripetal acceleration

If a body is rotating in one plane with a constant angular velocity about a fixed point, its linear velocity will always be tangential to the circle traced out by the body. The magnitude of the linear velocity will be constant. Thus

$$v = r\omega$$

But the velocity v will be in a continuously changing direction, as may be seen from figure 5.2.

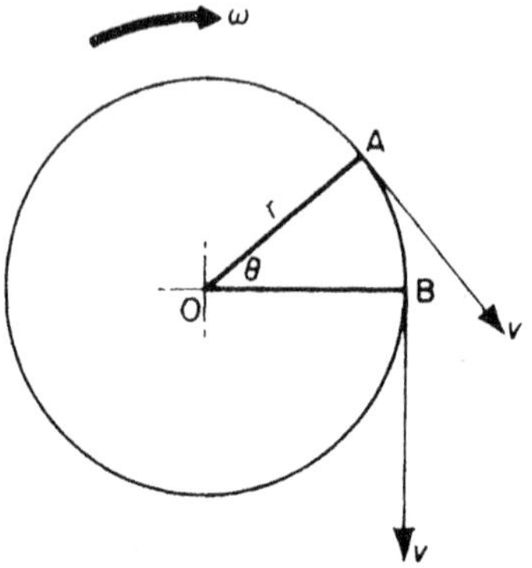

Figure 5.2 Centripetal acceleration

Now a velocity is only constant if both its magnitude and direction remain unchanged. Therefore in this case, even though the magnitude of the velocity is constant, the rotating body must be accelerating.

Suppose that the body is rotating clockwise, and is at a position A. Its linear velocity is then perpendicular to the radius OA and in a forward direction. Assume that after a short interval of t seconds the body has moved to a position B, where its linear velocity is perpendicular to OB. A velocity diagram (figure 5.3) can be drawn, to make

$$oa = ob = v$$

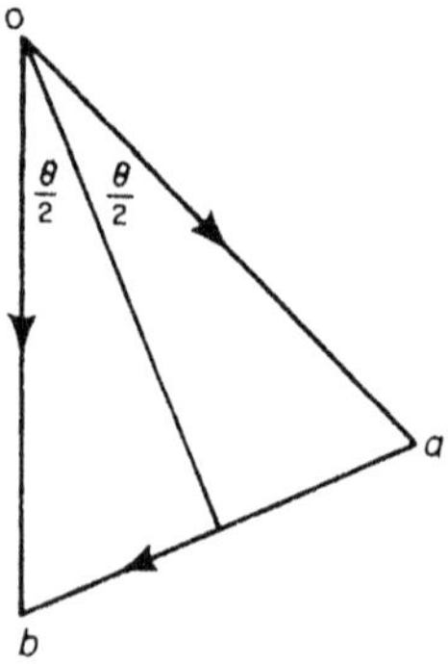

Figure 5.3 The triangle of velocities

where oa is perpendicular to OA and ob is perpendicular to OB.

The vector ab in figure 5.3 represents the velocity which must be added vectorially to the velocity represented by oa to give the velocity represented by *ob*. That is, *ab* represents the change in linear velocity of the body as it rotates from A to B.

Then from the triangle of velocities,

$$ab = 2v \sin \frac{\theta}{2}$$

If θ is made very small, A and B come closer together, and sin θ becomes very nearly equal to θ radians. Then

$$ab = 2v \times \frac{\theta}{2}$$

$$= v\theta$$

Thus the change in

$$\text{Linear velocity in time } t = v\theta$$

and the change in

$$\text{Linear velocity per second} = \frac{v\theta}{t}$$

But $\quad \theta = \text{Angular displacement} = \omega t$

$$\text{or,} \quad \omega = \frac{\theta}{t}$$

so that change of

$$\text{Linear velocity per second} = v \times \omega$$

$$= \frac{v^2}{r}, \text{ since } \omega = \frac{v}{r}$$

This means that the linear acceleration a at any point on the circumference is given by

$$a = \frac{v^2}{r} \qquad 5.4$$

Referring back to the vector diagram, when θ is very small indeed, *ob* becomes very nearly parallel to *oa*. At the same time, *ab* becomes nearly perpendicular to *oa*. As the value of θ becomes very small indeed, *ab* becomes perpendicular to *oa*. That is, the vector representing the change in linear velocity (which is the linear acceleration) is perpendicular to the vector representing the linear velocity. In other words, the acceleration at A is along the radius in the direction from A towards the centre of rotation. It is therefore known as a *centripetal acceleration*.

Centripetal force

Any acceleration must be produced by a force. From Newton's second law, the acceleration takes place in the direction of the applied force; there must therefore be a force acting along the radius, towards the centre, to produce centripetal acceleration. This force is known as a *centripetal force.*

From previous theory (p. 35):

$$\text{Force} = \text{Mass} \times \text{Acceleration}$$

$$\text{Centripetal force} = \frac{W}{g} \times \frac{v^2}{r}$$

$$= \frac{W}{g}\frac{v^2}{r} \qquad 5.5$$

where W is the weight of the rotating body.

Since $v = r\omega$, we can substitute for v in the above expression, to obtain

$$\text{Centripetal force} = \frac{W}{g}\,\omega^2 r \qquad 5.6$$

It should be noted that this force is always associated with the motion of a body in a circular path.

Centrifugal force

There must be another force acting at the centre of rotation, equal and opposite to the centripetal force. If this were not so the point O would move in the direction of the centripetal force. This opposing force, which is a reaction, is known as a *centrifugal force.*

It should be noted that it only exists as an equal and opposite reaction to the centripetal force.

Example

A wheel weighing 500 lbf is rotating at a uniform speed of 250 rev/min. Its centre of gravity is $\frac{1}{4}$ in. from the axis of rotation. Calculate the magnitude of the out-of-balance force.

Solution

$$\text{Out-of-balance force} = \text{Centrifugal force}$$

$$= \frac{W}{g}\,\omega^2 r \qquad \omega = \frac{\pi}{30} \times 250 \text{ rad/s}$$

$$= \frac{500}{32} \times \left(\frac{\pi \times 25}{3}\right)^2 \times \frac{1}{48}; \quad r = \frac{1}{4 \times 12}\text{ ft}$$

$$= 223 \text{ lbf}$$

Example

A car wheel, diameter 64 cm, is balanced by the application of a lead weight of 60 g at the periphery of the wheel-rim, which is 36 cm in diameter. What was the out-of-balance force on the wheel at a speed of 100 km/h before balancing?

Solution

$$\text{Wheel radius} = 0{\cdot}32\ \text{m}$$

$$\text{Wheel velocity} = \frac{100{,}000}{3{,}600} = 27{\cdot}73\ \text{m/s}$$

$$\text{Wheel angular velocity } \omega = \frac{v}{r}$$

$$= \frac{27{\cdot}73}{0{\cdot}32} = 86{\cdot}8\ \text{rad/s}$$

$$\text{Centripetal acceleration of weight} = \omega^2 r$$

$$= 86{\cdot}8 \times 86{\cdot}8 \times 0{\cdot}18\ \text{m/s}^2$$

$$= 1{,}352\ \text{m/s}^2$$

$$\text{Out-of-balance force} = \text{Mass} \times \text{Centripetal acceleration}$$

$$= 0{\cdot}06\ \text{kg} \times 1{,}352\ \text{m/s}^2$$

$$= 81{\cdot}12\ \text{N}$$

Summary

(*Note:* In all the following expressions ω, the angular velocity, is measured in radians per second.)

Kinetic energy of rotation

A rotating body possesses energy owing to its motion of rotation, quite separately from any energy of translation it may have. This energy is called its 'kinetic energy of rotation'.

The mass of the rotating body can be considered to be concentrated at some radius k from the axis or rotation: this radius is known as the 'radius of gyration' of the body.

In the same way as kinetic energy of translation is ½ mass × (velocity)^2, the kinetic energy of rotation is given by ½ mass × (velocity of mass concentration)^2, i.e.

$$\text{Kinetic energy of rotation} = \frac{1}{2}\frac{W}{g}(\omega k)^2 \quad \text{or} \quad \tfrac{1}{2}m(\omega k)^2$$

Moment of inertia

The quantity $\frac{W}{g} k^2$ is constant for any particular body rotating about a given axis. It is called the 'moment of inertia' of the mass about that axis of rotation, and is denoted by the symbol I.

$$\text{Moment of inertia } I = \text{Mass} \left(\frac{W}{g}\right) \times (\text{Radius of gyration } k)^2$$

$$I = \frac{W}{g} k^2 \quad \text{or} \quad mk^2$$

Using this expression, the kinetic energy of rotation can be written

$$\text{K.E. of rotation} = \tfrac{1}{2} I \omega^2$$

The flywheel

The purpose of a flywheel is to act as a reservoir of energy in a physical cycle which has intermittent, or variable, input and output of energy.

Centripetal acceleration

When bodies move in curved paths, the acceleration towards the centre of curvature is called a 'centripetal' acceleration. From the work of this chapter,

$$\text{Centripetal acceleration} = \frac{v^2}{r} = \omega^2 r$$

Centrifugal force

Centripetal action causes the curved path to be followed by the body. The reaction to this action is called 'centrifugal force.' Note that centrifugal force is a secondary, not a primary, phenomenon.

Exercises

1000 kg = 1 metric tonne = 1 t (mass)

1. A flywheel weighs 5000 lbf, and its mass may be assumed concentrated at the rim, which is at a radius of 4 ft from the centre of rotation. Calculate the kinetic energy of rotation of the flywheel at a speed of 140 rev/min.

2. A flywheel is 6 t mass and its mass may be assumed concentrated at the rim, which has a mean diameter of 2 m; Find (*a*) the kinetic energy of the flywheel when it is rotating at 150 rev/min, and (*b*) the loss of kinetic energy when the speed drops to 50 rev/min.

3. The centres of the crankpin and the crank shaft of an engine are 40 cm apart, and the crank shaft is rotating at 180 rev/min. Determine the centripetal acceleration of the crank pin.

4. The rim of a flywheel has a mean diameter of 6 ft and is rotating at 350 rev/min. What is the acceleration of a point on the rim.

5. A large gear wheel weighing $\frac{1}{4}$ tonf has its centre of gravity $\frac{1}{4}$ in. from the geometrical axis of the wheel. It is secured to a shaft and rotated at 300 rev/min. Calculate the magnitude of the out-of-balance force.

6. A motor-cyclist rides round a wall of death which has a diameter of 30 m. If the mass of the rider and his machine is 200 kg and the force between the wall and the motor-cycle is 4905 N, calculate the linear speed of the machine.

7. A car of mass 1 t travels at 65 km/h round a circular track of 40 m radius. Calculate the side thrust on the tyres. (The side thrust on the tyres is equal to the centripetal force required to keep the car moving in its circular track.)

8. An aircraft pulls out of a dive in a circular path of radius 2000 ft, whilst travelling at 500 knots. If the pilot weighs 200 lbf, what is the value of the force pressing him into his seat?

9. A motor-car, weighing 1 tonf travels over a hump-backed bridge in a road. If the radius of the arch is 60 ft, calculate (*a*) the force tending to lift the car off the road at a speed of 20 mile/h, and (*b*) the speed at which the wheels will just leave the road.

10. A rotor of mass 2000 kg is mounted centrally on a horizontal shaft which is supported on two bearings. If the centre of gravity of the rotor is displaced by $\frac{1}{10}$ cm from the central axis of the shaft, find the maximum and minimum load on each bearing when the rotor revolves at 560 rev/min.

11. A train weighing 320 t is travelling at 90 km/h round a circular portion of a track of radius 5000 km. What is the total thrust exerted by the rails on the wheel flanges?

12. A rotor weighing 1000 lbf is carried on a shaft mounted vertically between two bearings. If the centre of gravity of the rotor is displaced $\frac{1}{8}$ in. from the central axis of the shaft, find the total force on the bearings when the rotor rotates at 500 rev/min.

6

Friction and the inclined plane

All surfaces, even although they may appear smooth to the naked eye, consist of a very large number of irregular shapes which, when magnified under a microscope, cause the surfaces to appear rough.

If two such surfaces are brought into contact with each other, and a force is applied tending to make one surface slide over the other, it will be found that the 'rough' surfaces resist motion. This resistance is known as 'friction', and the force which must be applied to overcome the resistance is known as the 'force of friction'.

If we make an experiment with two rough surfaces which are dry (that is, no form of lubrication is applied) we shall find that the force required to start one surface sliding over the other is greater than that required to keep the surface sliding at a constant velocity. The resistance to the force required to start motion is called 'static friction,' whereas the smaller resistance to the force required to maintain motion at a steady speed is called 'kinetic friction'.

Limiting friction

Suppose a rectangular block of material is resting on a flat, horizontal surface, and the two surfaces in contact are smooth to the eye but rough when viewed under a microscope. If a small force P is applied

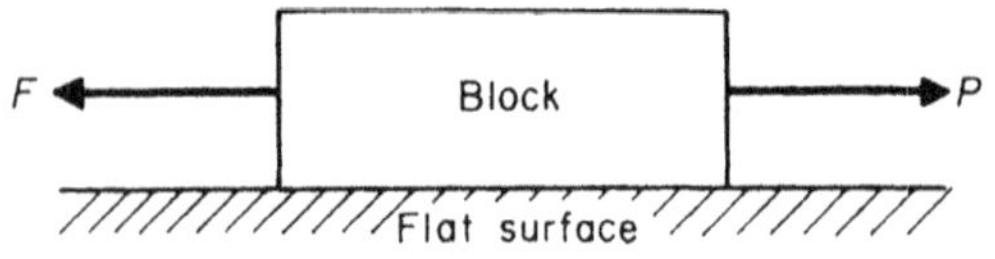

Figure 6.1 Resisted motion

to the block its movement will be resisted by the force of friction F, (figure 6.1). If no movement takes place the force P will be balanced by an equal and opposite force F. Further, the weight W of the block, acting vertically downwards, will be resisted by the flat surface. Thus, a reaction R will act upwards from the flat surface as in figure 6.2.

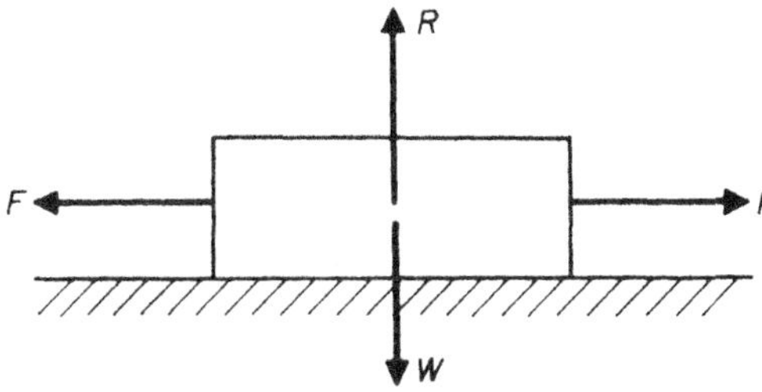

Figure 6.2 Condition for equilibrium, block stationary.

These four forces will be in equilibrium, and so, if no movement of the block occurs, we can say

$$R = W \quad \text{and} \quad F = P$$

Now let the force P be increased slowly. The frictional force F will also increase to maintain the balance. However, although we can go on increasing the applied force P, the value of the frictional force cannot increase indefinitely. Instead, the value of F will approach a certain maximum value. Up to this maximum value F will balance P and no motion will occur.

If, at this value of P (corresponding to the maximum value of F) the block were set in motion, it would continue to move with a uniform velocity provided the resistance between the two surfaces remained unchanged.

Any increase in this value of P, or any reduction in the resistance between the two surfaces, would permit the block to accelerate in the direction of the applied force.

The maximum value of F, when motion is about to start, is called 'the limiting friction'.

The laws of friction for dry surfaces

There are four laws governing the friction of dry, or unlubricated, surfaces. These laws apply to bodies which are at rest, or which are about to start uniform motion, and are only approximately true. In this book we shall not consider accelerated motion.

The laws can be stated thus:

1. Friction opposes motion. That is, if one surface is sliding over

another, or is tending to slide over another, then the frictional force is acting in a direction opposite to that in which motion is occurring, or trying to occur.

The law does NOT imply that the resistance offered by the rough surfaces, or the frictional force, acts in a direction immediately opposite to that of the APPLIED FORCE (see figure 6.3).

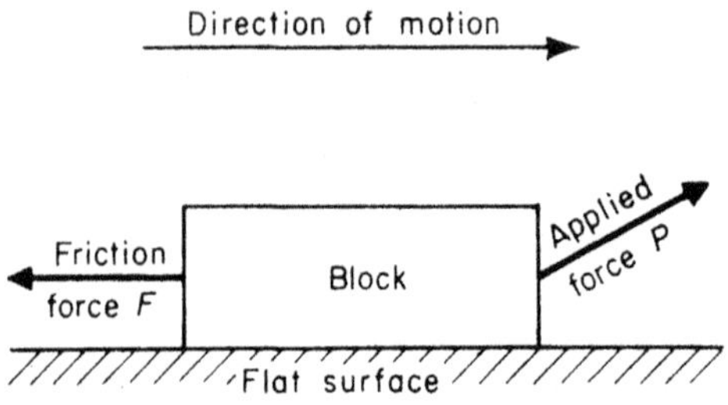

Figure 6.3 Friction opposes motion.

2. The force of friction depends upon the nature of the surfaces in contact. That is, the resistance between the two surfaces in contact depends on the materials involved, and the roughness of the surfaces.

3. When motion is about to take place the magnitude of the limiting friction is directly proportional to the NORMAL reaction between the two surfaces in contact. This reaction is vertical ONLY when the two surfaces are horizontal.

4. The force of friction is independent of the areas in contact. For a given load, a larger area results in a smaller intensity of loading. The force of friction, however, is not dependent on the intensity of loading, but only on the total load.

Coefficient of friction

(symbol μ—Greek letter mu)

When motion is about to take place, or, when the block is moving with a uniform velocity over the rough surface, the magnitude of the frictional force F is directly proportional to the normal reaction R.

In other words, the ratio

$$\frac{F}{R} = \text{Constant}$$

This constant is known as the "coefficient of friction" of the two surfaces in contact.

The value of the constant depends—as stated in the second and third laws of friction—on the nature of the two surfaces in contact, and also on the normal reaction R. The magnitude of F will change if only one of the surfaces is changed.

The symbol used for the coefficient of friction is the Greek letter μ, and so, on figure 6.4

$$\frac{F}{R} = \mu$$

or,

$$F = \mu R \qquad 6.1$$

Note that F and R are forces which the rough surface exert on the block.

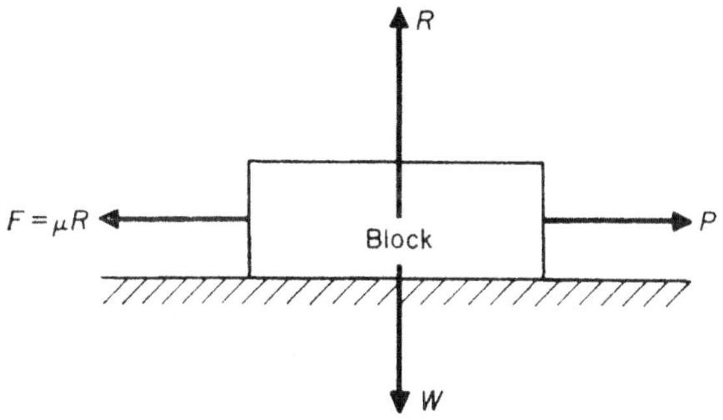

Figure 6.4 The coefficient of friction.

These two forces can be combined, if desired, into a single force S (as in figure 6.5) acting at an angle ϕ to the normal reaction.

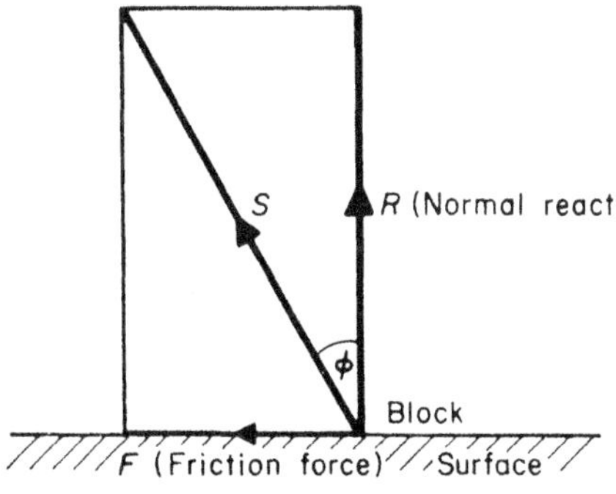

Figure 6.5 The angle of friction.

S is the resultant of the two forces R and F. We can say, from the triangle of forces

$$\frac{F}{R} = \tan \phi$$

But $$\frac{F}{R} = \mu$$

and so $$\frac{F}{R} = \mu = \tan\phi$$

or, $$\mu = \tan\phi \qquad 6.2$$

The angle ϕ is called 'the angle of friction'.

Friction and the inclined plane

Consider a block of weight W resting on a rough plane surface inclined at an angle θ to the horizontal, as in figure 6.6.

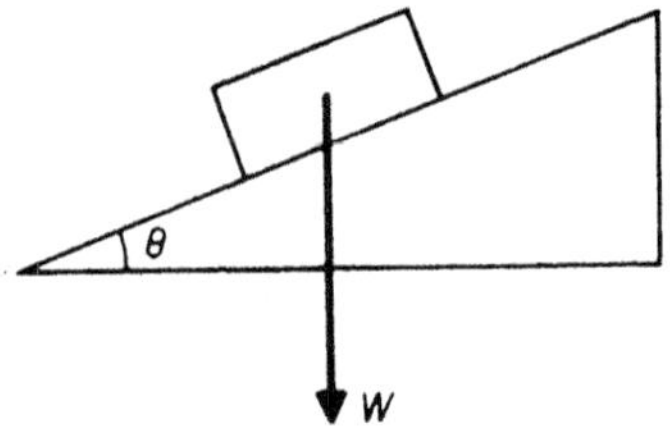

Figure 6.6 A block resting on an inclined surface.

In general we shall be dealing with four forces acting on the block, and maintaining it in equilibrium. These forces are:

1. Its weight always acting vertically downwards.
2. A reaction between the block and the inclined surface, acting normal to the surface.
3. A frictional force opposing motion and acting parallel to the inclined plane.
4. An external force applied to the block to produce the motion required.

If we want to establish relationships between these forces we must be quite sure that we can identify them, and so, if we can sketch correctly a simple diagram showing the directions of the forces concerned we have gone a long way to solving the problem.

Consider first of all the directions of the four forces concerned. Two of these can be drawn on a line diagram of an inclined plane without even examining the problem in detail. Thus, the weight (W) always acts vertically downwards, and the normal reaction (R) can be drawn upwards and at right angles to the inclined surface. Sketch these in at once, and add the two angles θ as shown in figure 6.7.

You will eventually need both of these angles.

Next consider the frictional force. Each problem will state in which direction the block is moving, or sliding, or 'tending to move'. The frictional force opposes this motion, and so the force F can be added to our diagram. Remember that the frictional force must be parallel to

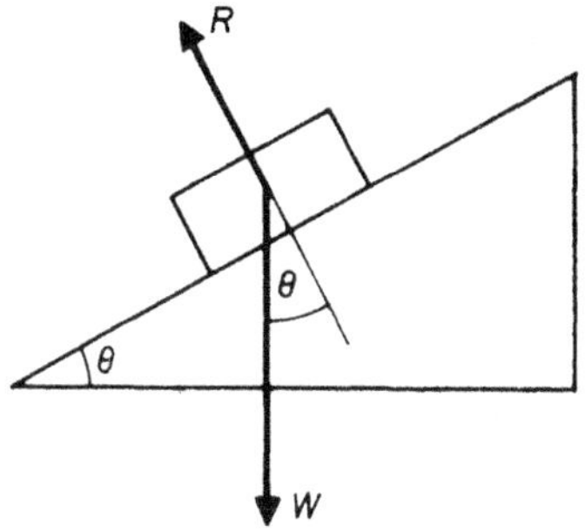

Figure 6.7 The weight and the normal reaction.

the inclined surface. Remember also that one of the laws of friction states that the value of F, when motion is about to take place, is proportional to the normal reaction, and so, as in equation 6.1

$$F = \mu R$$

If, therefore, the block is tending to slide DOWN the plane, the frictional force F will act UP the plane, as shown in figure 6.8(a).

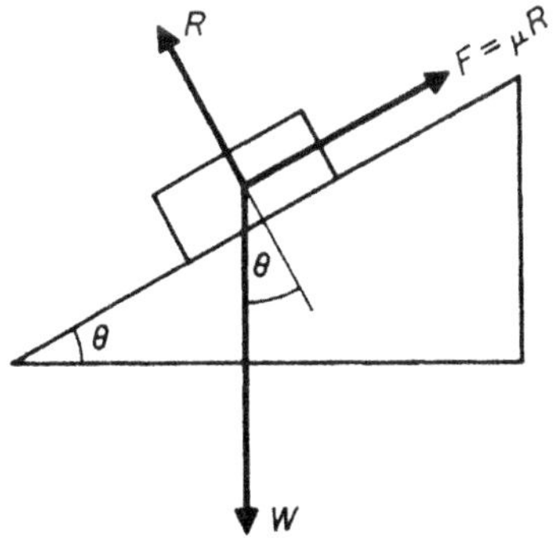

Figure 6.8(a) Block sliding down.

Conversely, if the block is being pulled, or pushed, UP the plane, the frictional force will act DOWN the plane, as shown in figure 6.8(b).

We are now left with the externally applied force P, which is giving the block its motion. The direction of this force will be stated in the problem, and can, theoretically, be inclined at any angle to the inclined plane. Usually, however, it is either parallel to the plane, or it is horizontal.

Should this applied force be given as a 'push' force it is advisable to convert it immediately into a 'pull' force.

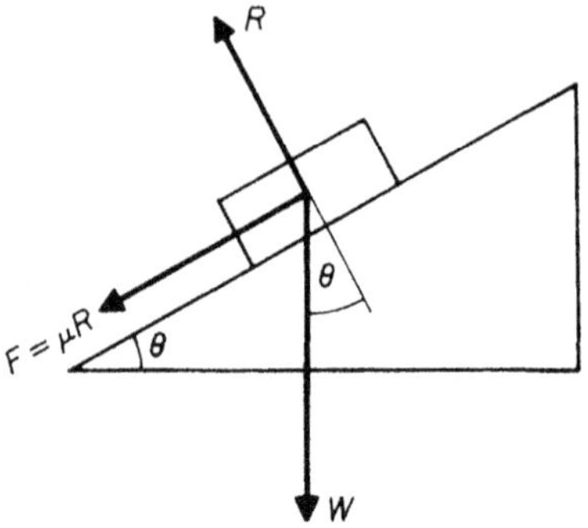

Figure 6.8(*b*) Block sliding up.

We are now in a position to start solving a problem. Remember that we are going to consider the equilibrium of the block under the action of 4 forces. Therefore, resolve these forces in two directions at right

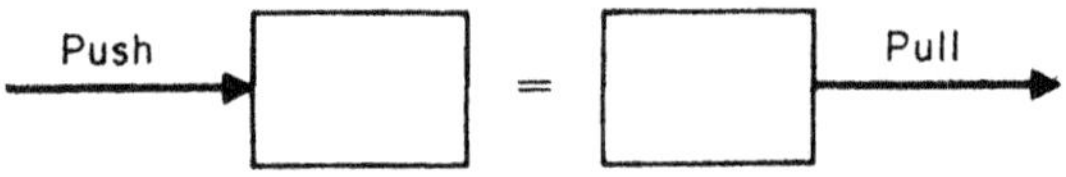

Figure 6.9 "Push" force to a "Pull" force.

angles, for in this way we shall obtain two equations which will enable us to determine our unknown quantity.

It is appropriate here to review some of the work which is normally done either in your G2 course or at your school.

The resultant of two perpendicular forces

Suppose that two forces P and Q acting at a single point O along directions which are at right angles to each other. The resultant can be determined either graphically or analytically.

The analytical method is much easier—mainly because we have selected two directions at right angles to each other—and is also more accurate.

Thus, in the triangle OBC, where CB = OA = P

$$R^2 = P^2 + Q^2$$

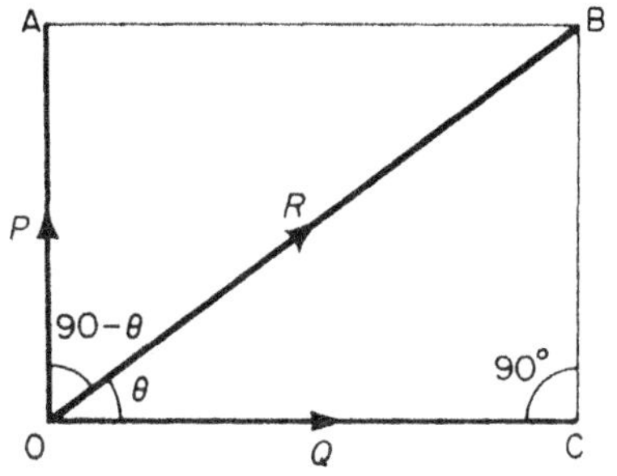

Figure 6.10 The resultant of 2 perpendicular forces.

or,
$$R = \sqrt{P^2 + Q^2}$$

and
$$\tan \theta = \frac{P}{Q}$$

giving us R and θ to any degree of accuracy we require.

Resolution of forces

Suppose now that we are given the values of the resultant R and the angle θ which this resultant makes with one of the two forces P and Q, and we are asked to determine these forces.

The process of determining the values of the two forces P and Q is called 'resolving' the force R, and the two forces P and Q are called the 'resolved parts,' or 'components' of R.

Note that we shall only concern ourselves with components which are acting at right angles to each other. By selecting directions perpendicular to each other we shall (*a*) simplify our calculations, and (*b*) ensure that each component has no effect in the direction of the other.

Again, we can, if we wish, determine the values of the components either graphically or analytically, the latter method being the simpler of the two.

Thus, referring to figure 6.10

$$\frac{\text{OC}}{\text{OB}} = \cos \theta \quad \text{and} \quad \frac{\text{OA}}{\text{OB}} = \cos (90 - \theta)$$

or,
$$\frac{Q}{R} = \cos \theta \quad \text{and} \quad \frac{P}{R} = \cos (90 - \theta)$$

$$\therefore\ Q = R \cos \theta \quad \text{and} \quad P = R \cos (90 - \theta)$$

Consequently, the component, or the resolved part, of R in the direction OC is $R \cos \theta$, and the component, or the resolved part, of R in the direction OA is $R \cos (90 - \theta)$. This leads to the general rule, namely, that 'The resolved part of a force in a given direction is equal to the product of the force and the cosine of the angle between the direction of the force and the direction chosen'.

When we consider two directions at right angles, such as OX and OY in figure 6.11, it so happens that cos $(90 - \theta)$ has the same value

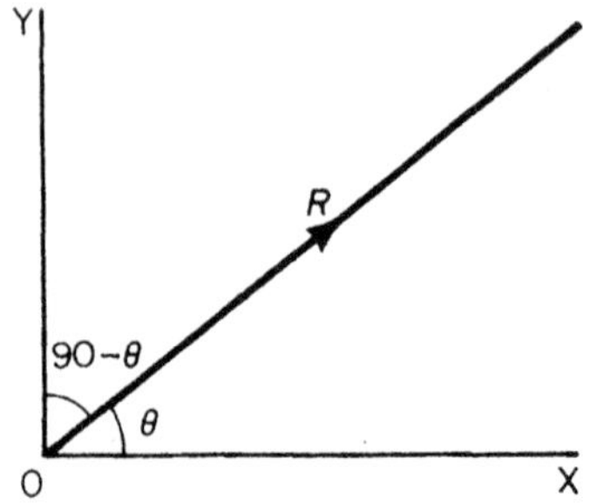

Figure 6.11 The resolved parts of a force.

as sin θ, but do not let this fact interfere with your remembering the basic rule of resolving, namely:

> The resolved part of a force in a given direction = magnitude of force × cos (angle between the direction of the force and the given direction).

In due course, because you will always be considering two directions which are at right angles to each other, it will become automatic to substitute sin θ for cos $(90 - \theta)$.

Resolved part as a projected length

The resolved part of a force R can also be considered in terms of a projected length. Thus, if OB, in figure 6.12(*a*), represents to scale the force R in a direction inclined at angle θ to the horizontal, OC is the projected length of OB in the horizontal direction, and so represents to scale the resolved part of R in the horizontal direction.

Similarly, in figure 6.12(*b*), OA is the projected length of OB in a vertical direction, and so represents to scale the resolved part of R in the vertical direction.

Returning now to the four forces acting on the block which is resting on an inclined plane, it does not usually matter which two directions are selected for the resolution of the forces, but the work will be greatly simplified if you select one of two pairs of directions. These are:

1. Horizontal and vertical directions; or
2. Parallel and perpendicular to the inclined plane.

Take some possible problems.

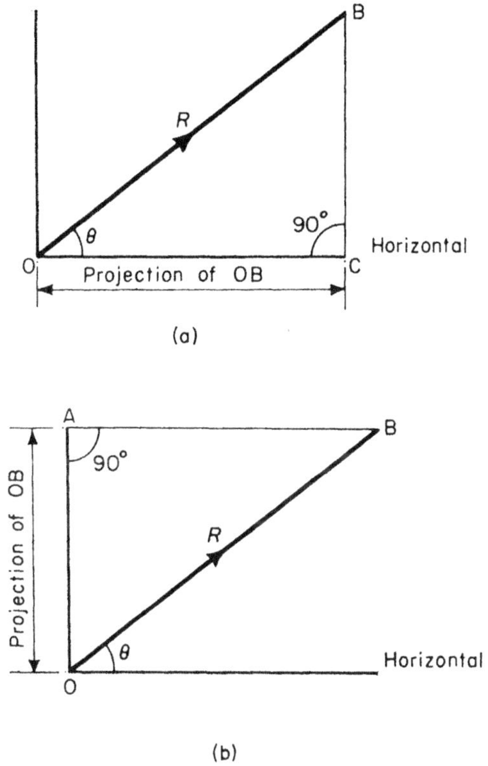

Figure 6.12(*a*) and (*b*) Resolved parts as projected lengths.

Problem 1

(*a*) Block moving UP the plane.
(*b*) Force applied parallel to the inclined surface.

Procedure

1. Draw a sketch of

(*a*) the surface inclined at an angle θ to the horizontal, and
(*b*) the block resting on the surface.

2. Add weight W (figure 6.13) acting vertically downwards, and R normal to the inclined surface, making an angle θ with the vertical.

Because the block is moving UP the plane we can

3. add μR acting down the plane
4. add force P acting upwards, and, as required, parallel to the inclined surface.

We have thus introduced ALL the forces—a total of 4.

Note that 3 of them (μR, R and P) are in directions parallel and perpendicular to the plane. Because of this we shall have no hesitation in selecting these 2 directions as our reference lines, and in resolving all the forces in these directions.

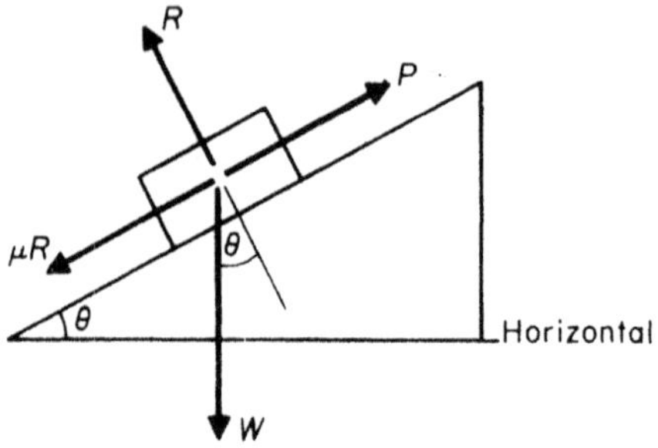

Figure 6.13 Block moving up the plane.

We have seen that it is only necessary to evaluate the resolved part of one force—W—in the two directions selected.

Thus, resolved part of W in a direction parallel to the plane is

$$W \cos (90 - \theta) = W \sin \theta \text{ down the plane}$$

Also, resolved part of W in a direction perpendicular to the plane is

$$W \cos \theta \text{ opposing the force } R$$

We may, if we wish, replace the single force W by these two components and re-draw the diagram as in figure 6.14.

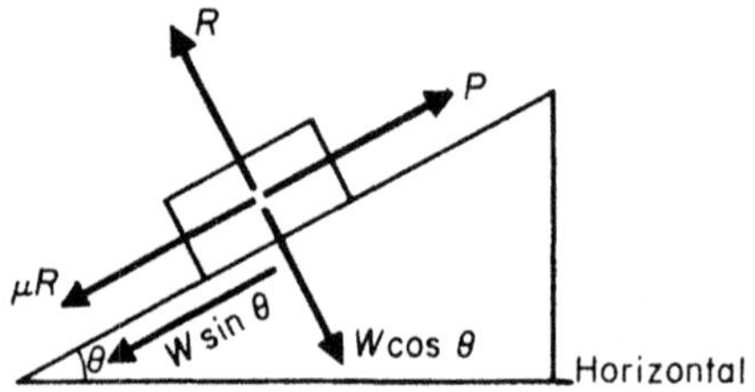

Figure 6.14 The resolved forces. Block moving up.

For equilibrium we must have two equations representing forces resolved in the two selected directions.

Thus, in a direction parallel to the plane,

Total upward forces = Total downward forces

$$\therefore \; P = \mu R + W \sin \theta \qquad 6.3$$

and, in a direction perpendicular to the plane,

Total outward forces = Total inward forces

$$\therefore \; R = W \cos \theta \qquad 6.4$$

In most problems we shall not need to evaluate R, and so we can combine equations 6.3 and 6.4 with the aim of eliminating R.

Thus, by substituting the value of R (equation 6.4) in equation 6.3, we get

$$P = W(\mu \cos\theta + \sin\theta) \qquad 6.5$$

There is no need to remember this equation. Always proceed from first principles and you will soon find yourself able to solve any problem of a similar nature.

If you have decided to use the form

$$\cos(90 - \theta) = \sin\theta$$

there is one golden rule to remember as a final check on your two equations, (equations 6.3 and 6.4.).

Rule

If one equation contains a term involving a force multiplied by $\cos\theta$, then the other equation MUST also contain a similar term with the same force multiplied by $\sin\theta$.

Problem 2

(*a*) Block moving UP the plane.
(*b*) 'Push' force in a horizontal direction.

Procedure

1. Draw a sketch of

(*a*) the surface inclined at an angle θ to the horizontal, and
(*b*) the block resting on the surface.

2. Add weight W acting vertically downwards, and R normal to the inclined surface making an angle θ with the vertical.

Because the block is moving UP the plane we can

3. Add μR acting down the plane.

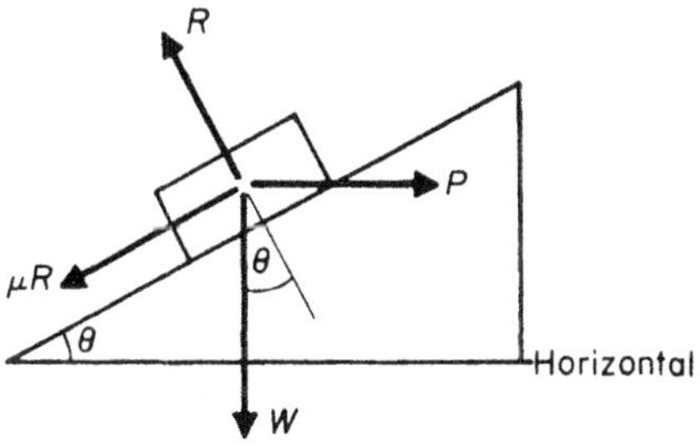

Figure 6.15 Block moving up. Force horizontal.

4. Convert force P into a 'pull' force, and then add it, acting in a horizontal direction.

We have thus introduced ALL the forces—a total of 4.

Note that 2 of them —P and W—are in horizontal and vertical directions, whereas μR and R are in directions parallel and perpendicular to the inclined surface. We shall thus be left to resolve 2 forces no matter which two directions are selected as our reference lines. There is, therefore, little preference to be given to either choice of directions.

We may re-sketch our diagram, if desired, as in figure 6.16, where we have replaced:

1. R acting outwards by $R \cos \theta$ acting vertically upwards and $R \sin \theta$ acting horizontally.

2. μR acting down the plane by $\mu R \cos \theta$ acting horizontally and $\mu R \sin \theta$ acting vertically downwards.

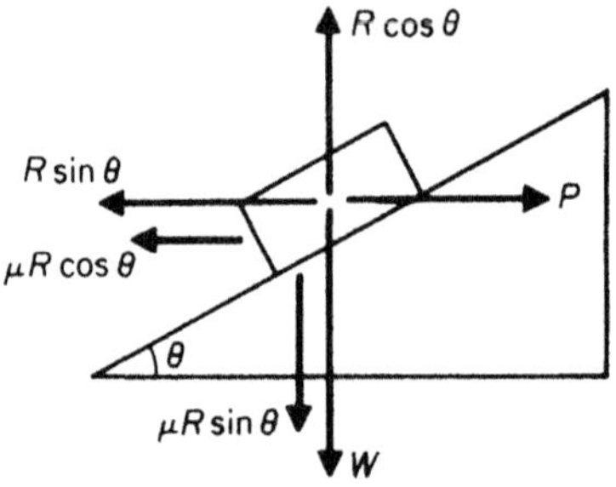

Figure 6.16 Resolved forces. Block moving up. Force horizontal.

For equilibrium,

1. Total downward force = Total upward force

$$W + \mu R \sin \theta = R \cos \theta$$

or,

$$W = R \cos \theta - \mu R \sin \theta$$

$$= R(\cos \theta - \mu \sin \theta) \qquad 6.5$$

and

2. Total inward force = Total outward force

$$P = R \sin \theta + \mu R \cos \theta$$

$$= R(\sin \theta + \mu \cos \theta) \qquad 6.6$$

We can then eliminate R from equations 6.5 and 6.6. by division, thus

$$\frac{P}{W} = \frac{\sin \theta + \mu \cos \theta}{\cos \theta - \mu \sin \theta}$$

Again, there is no need to memorize the formula, although this particular one can, if desired, be put into a more simple form.

Problem 3

(*a*) Block moving DOWN the plane.
(*b*) 'Push' force parallel to the inclined surface, and up the plane.

Procedure

1. Draw a sketch of

(*a*) the surface inclined at an angle θ to the horizontal, and
(*b*) the block resting on the surface.

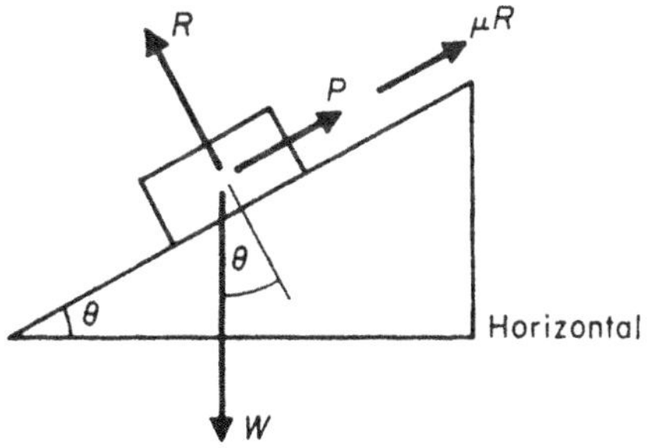

Figure 6.17 Block moving down. Push force up the plane.

2. Add weight W acting vertically downwards, and R normal to the inclined surface making an angle θ with the vertical.

Because the block is moving DOWN the plane we can

3. add μR acting up the plane
4. convert force P into a 'pull' force, and then add it acting upwards and parallel to the inclined surface.

We have thus introduced ALL the forces—a total of 4.

Note that 3 of them (P, μR and R) are in directions parallel to and perpendicular to the plane. Because of this we shall select these 2 directions as our reference lines, and resolve all the forces in these directions.

We can replace the single force W by two components in the selected directions, and re-draw the diagram (figure 6.18).

For equilibrium,

1. Total upward force = Total downward force

$$P + \mu R = W \sin \theta \qquad 6.7$$

and

2. Total outward force = Total inward force

$$R = W \cos \theta \qquad 6.8$$

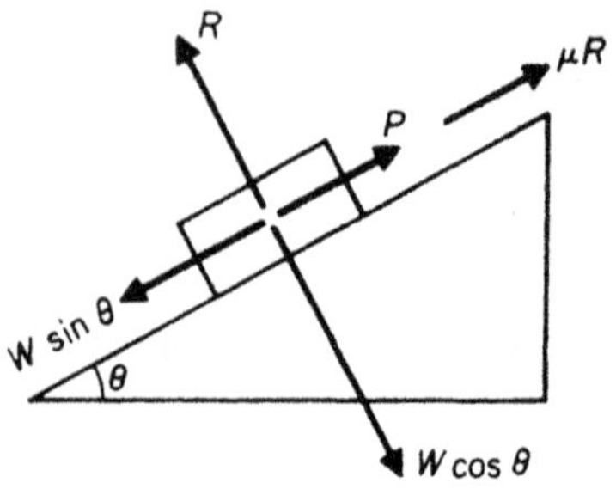

Figure 6.18 Resolved forces. Block moving down. Push force up the plane.

We can eliminate R by combining equations 6.7 and 6.8. Thus,

$$P + \mu(W\cos\theta) = W\sin\theta$$

$$P = W\sin\theta - \mu W\cos\theta$$

$$= W(\sin\theta - \mu\cos\theta)$$

There is no need to memorize this equation.

Problem 4

(*a*) Block moving DOWN the plane.
(*b*) 'Push' force acting in a horizontal direction.

Procedure

1. Draw a sketch of

(*a*) the surface inclined at an angle θ to the horizontal, and
(*b*) the block resting on the surface.

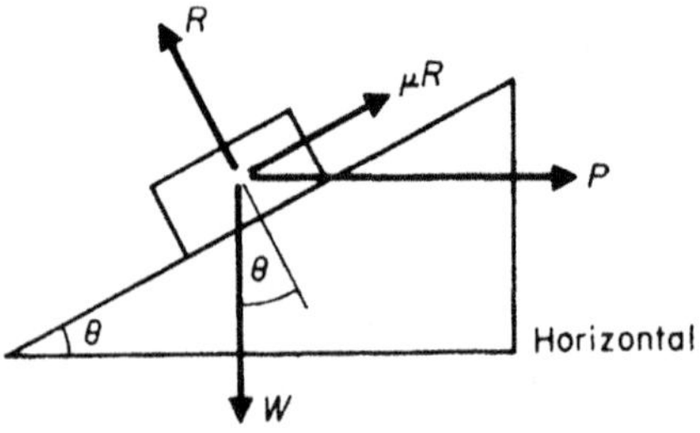

Figure 6.19 Block moving down. Push force horizontal.

2. Add the weight W acting vertically downwards, and R normal to the inclined surface making an angle θ with the vertical.

Because the block is moving DOWN the plane we can

3. add μR acting up the plane

4. convert force P into a 'pull' force, and then add it, acting in a horizontal direction.

We have thus introduced ALL the forces—a total of 4.

Note that 2 of them—P and W—are in horizontal and vertical directions, whereas μR and R are in directions parallel and perpendicular to the inclined surface.

As in problem 2 there is little to choose when selecting the two directions as reference lines. Let us again select the horizontal and vertical directions.

We may then re-sketch the diagram, if desired.

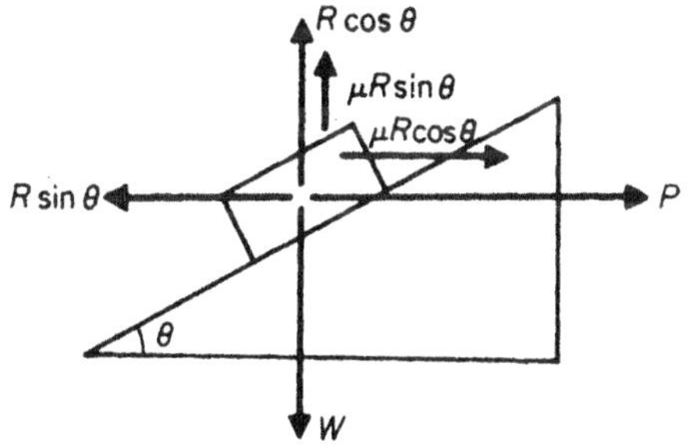

Figure 6.20 Resolved forces. Block moving down. Push force horizontal.

For equilibrium,

1. Total downward force = Total upward force

$$W = R\cos\theta + \mu R\sin\theta$$
$$= R(\cos\theta + \mu\sin\theta) \qquad 6.9$$

and

2. Total inward force = Total outward force

$$P + \mu R\cos\theta = R\sin\theta$$
$$P = R\sin\theta - \mu R\cos\theta$$
$$= R(\sin\theta - \mu\cos\theta) \qquad 6.10$$

We can eliminate R by dividing equation 6.10 by equation 6.9, thus

$$\frac{P}{W} = \frac{\sin\theta - \mu\cos\theta}{\cos\theta + \mu\sin\theta}$$

As in problem 2, this expression can, if desired, be put into a simple form, but there is no justification for doing so, as it need not be memorized.

Example

A rectangular block of iron, weighing 200 N, rests on a rough horizontal floor. If the coefficient of friction between the metal and the floor is 0·5, calculate the force required to drag the block along the floor with uniform velocity when (*a*) the force is applied horizontally, and (*b*) the force acts at an angle of 30° upwards from the horizontal.

Solution

(*a*)

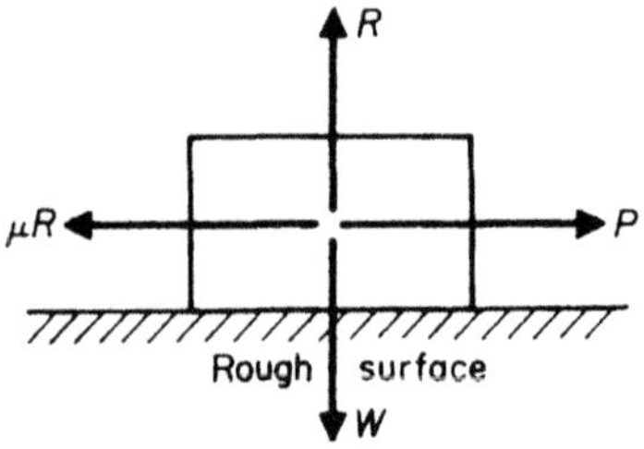

Figure 6.21 Applied force horizontal.

Equating vertical forces

$$R = W$$

Equating horizontal forces

$$\mu R = P$$

Eliminating R from these two equations, we get

$$\begin{aligned} P &= \mu W \\ &= 0{\cdot}5 \times 200 \\ \therefore \text{ Applied force} &= 100 \text{ N} \end{aligned}$$

(*b*)

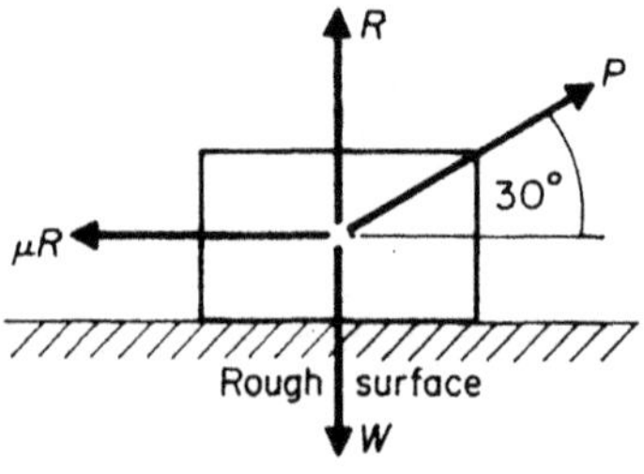

Figure 6.22 Force at 30° to horizontal.

Resolving forces horizontally

$$\mu R = P \cos 30^\circ$$

Resolving forces vertically

$$R + P \sin 30° = W$$

Eliminating R from these two equations,

$$\mu(W - P \sin 30°) = P \cos 30°$$

Substituting known values

$$0{\cdot}5(200 - 0{\cdot}5P) = 0{\cdot}866P$$
$$100 - 0{\cdot}25P = 0{\cdot}866P$$
$$100 = 1{\cdot}116P$$

or, $$P = 89{\cdot}7 \text{ lbf}$$

Example

A block of steel, of mass 500 kg, rests on a rough surface inclined at 10° to the horizontal. If the coefficient of sliding friction between the block and the surface is 0·25 find the magnitude of the force which, when applied parallel to the inclined surface, is necessary (*a*) to push the block up the incline, and (*b*) to push the block down the incline.

(*a*)

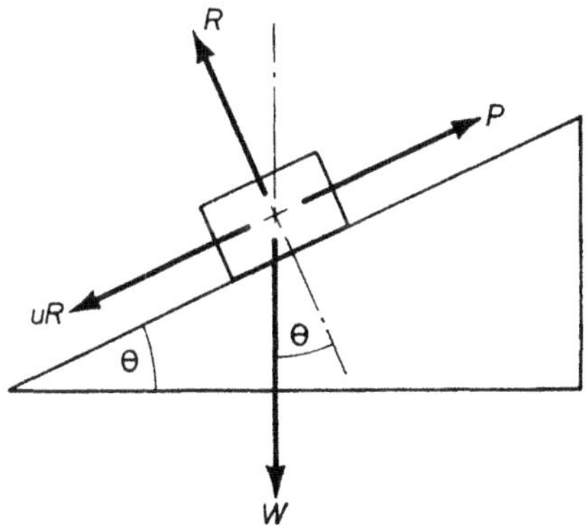

Figure 6.23 Force applied up the incline.

First, replace a 'push' force by a 'pull' force P. Remember, friction opposes motion and equals μ times the normal reaction.

Resolving forces parallel to the plane

$$P = \mu R + W \sin \theta$$

Resolving forces perpendicular to the plane

$$R = W \cos \theta$$

Eliminating R from these two equations, we get

$$P = \mu W \cos \theta + W \sin \theta$$
$$= W(\mu \cos \theta + \sin \theta)$$

Substituting known values

$$P = 500(0{\cdot}25 \times 0{\cdot}9848 + 0{\cdot}1736)$$
$$= 500(0{\cdot}2462 + 0{\cdot}1736)$$
$$= 500 \times 0{\cdot}4198$$
$$= 209{\cdot}9 \times 9{\cdot}81 \text{ N/kg}$$

(*b*) $$= 2059 \text{ N}$$

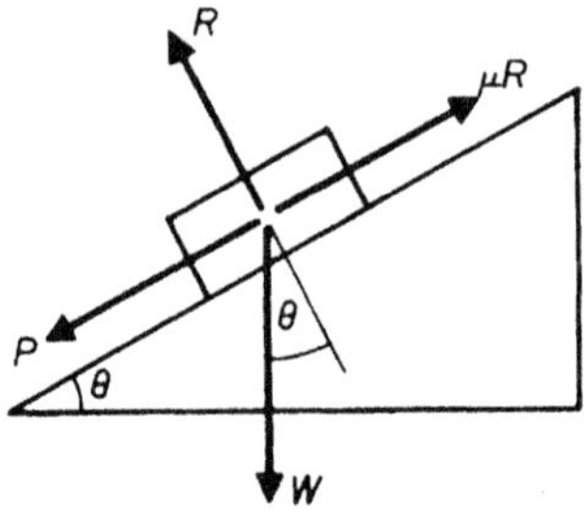

Figure 6.24 Force applied down the incline.

First, replace a 'push' force by a 'pull' force P. The frictional force, opposing motion, will this time act up the inclined plane.

Resolving the forces parallel to the plane

$$\mu R = P + W \sin\theta$$

Resolving forces perpendicular to the plane

$$R = W \cos\theta$$

Eliminating R from these two equations, we get

$$\mu W \cos\theta = P + W \sin\theta$$

or $$P = W(\mu \cos\theta - \sin\theta)$$

Substituting known values

$$P = 500(0{\cdot}25 \times 0{\cdot}9848 - 0{\cdot}1736)$$
$$= 500(0{\cdot}2462 - 0{\cdot}1736)$$
$$= 500 \times 0{\cdot}0726$$
$$= 36{\cdot}3 \text{ kg} \times 9{\cdot}81 \text{ N/kg}$$
$$= 356 \text{ N}$$

Example

What horizontal force is required to push a block of steel weighing 100 lbf up a slope inclined at 30° to the horizontal, if the coefficient of friction between the block and the inclined surface is 0·25.

Solution

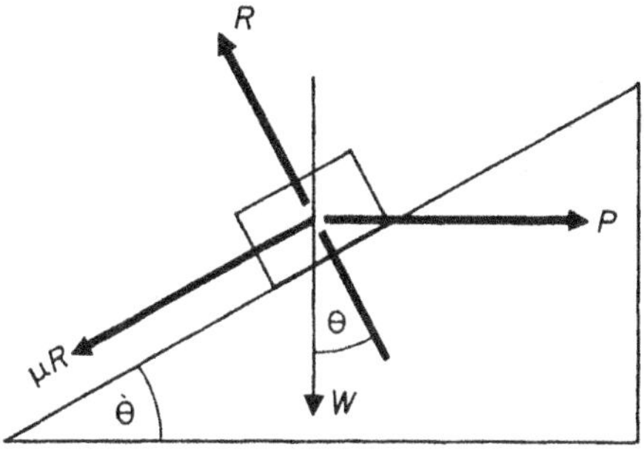

Figure 6.25 Horizontal force.

First replace the 'push' force by a 'pull' force P.
The frictional force, opposing motion, will act down the slope.
Resolving forces parallel to the plane

$$P \cos \theta = \mu R + W \sin \theta$$

Resolving forces perpendicular to the plane

$$R = P \sin \theta + W \cos \theta$$

Eliminating R from these two equations, we get

$$P \cos \theta = \mu P \sin \theta + \mu W \cos \theta + W \sin \theta$$

or,

$$P(\cos \theta - \mu \sin \theta) = W(\mu \cos \theta + \sin \theta)$$

$$P = W \frac{\mu \cos \theta + \sin \theta}{\cos \theta - \mu \sin \theta}$$

Substituting known values,

$$P = 100 \left(\frac{0{\cdot}25 \times 0{\cdot}866 + 0{\cdot}5}{0{\cdot}866 - 0{\cdot}25 \times 0{\cdot}5}\right)$$

$$= 100 \left(\frac{0{\cdot}717}{0{\cdot}741}\right)$$

$$= 96{\cdot}8 \text{ N}$$

Angle of repose
Suppose a block, weighing W, rests on a horizontal rough surface, and the surface is tilted upwards slowly. There is a tendency for the block to slide down the inclined surface, but it is prevented from doing so by the frictional force, which, as we have seen, opposes motion.

Earlier in this chapter we saw that the frictional force can increase up to a maximum value, known as 'the limiting value', which is reached when motion begins to take place.

Therefore, as the angle of the plane is increased, so the value of the frictional force F increases—to balance the increased effect of W—until at a certain angle, the block begins to slide down the plane.

The value of β in figure 6.26, when motion is about to take place, is called the ANGLE OF REPOSE. It has the same value as the friction angle ϕ.

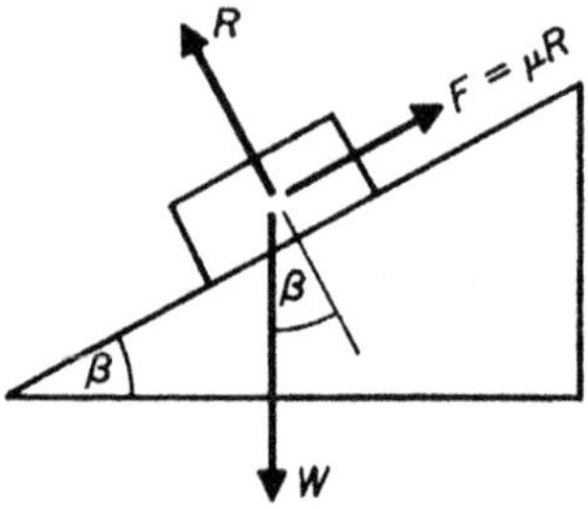

Figure 6.26 The angle of repose.

At this angle β the value of F is the limiting value of the frictional force, and this is equal to μR.

Hence, when β is the angle of respose

$$F = \mu R$$

and so

$$\frac{F}{R} = \mu = \tan\phi = \tan\beta$$

Example

A block of stone of mass 15 kg is on the point of slipping down a rough surface inclined upwards at an angle of 20° to the horizontal. Write down the angle of repose, and also the limiting coefficient of friction. What is the value of the frictional force.

Solution

Because the block is on the point of slipping down the incline under the influence of the gravitational force, the angle made by the rough surface with the horizontal is equal to the angle of repose. Hence,

$$\text{Angle of repose} = 20°$$

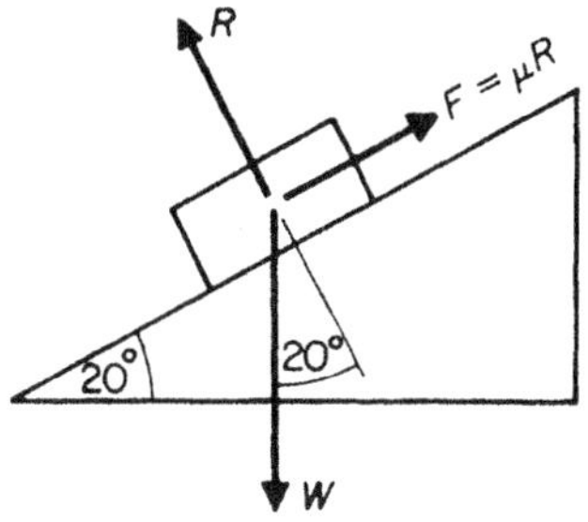

Figure 6.27 The angle of repose.

Because the block in figure 6.27 is in equilibrium under the influence of 3 forces—W, R and F—we can write down

$$R = W \cos \beta$$

But $$F = \mu R$$

where $$\mu = \tan \phi = \tan \beta$$
$$= \tan 20°$$

or, $$\mu = 0{\cdot}364$$

Also, $$R = 15 \cos 20°$$
$$= 15 \times 0{\cdot}9397 \times 9{\cdot}81$$

and, because $$F = \mu R$$
$$= 0{\cdot}364 \times 15 \times 0{\cdot}9397 \times 9{\cdot}81$$

The frictional force

$$F = 5{\cdot}13 \times 9{\cdot}81$$
$$= 50{\cdot}3 \text{ N}$$

when motion is about to begin.

Geometrical and Graphical solutions

We have seen that two of the forces mentioned in the previous work—the normal reaction R and the frictional force F—are constant for any particular set of conditions. They can, therefore, be combined if necessary into one frictional resultant force, which we have called S.

Geometrical and graphical solutions use this single resultant force to replace the normal reaction R and the frictional force F.

It is necessary to remember the following facts about the frictional resultant S:

(*a*) for limiting friction, it is always inclined at an angle ϕ ($= \tan^{-1} \mu$) to the normal to the plane.

(*b*) the direction of friction is such as to oppose motion between the surfaces; that is, S will always be drawn on the side of the normal remote from the direction of sliding.

(*c*) the block is in equilibrium under the action of three forces; its weight, the applied force, and the frictional resultant S.

The two figures 6.28(*a*) and (*b*) show the use of the vector triangle in solving friction problems of this kind. Applied forces P_p and P_h act separately.

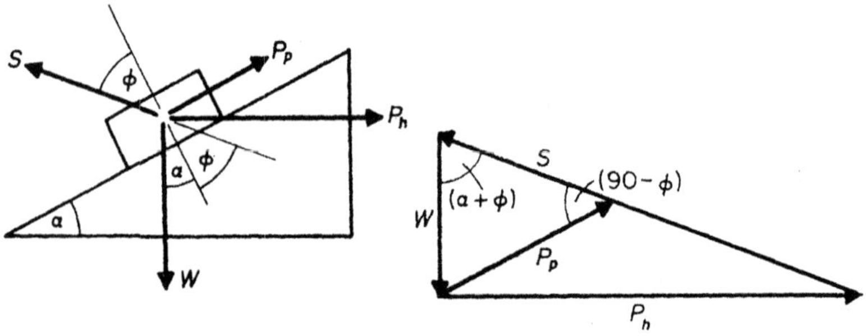

(a) Force and motion UP plane

(a)

Figure 6.28(*a*) Applied force tending to cause motion up the plane.

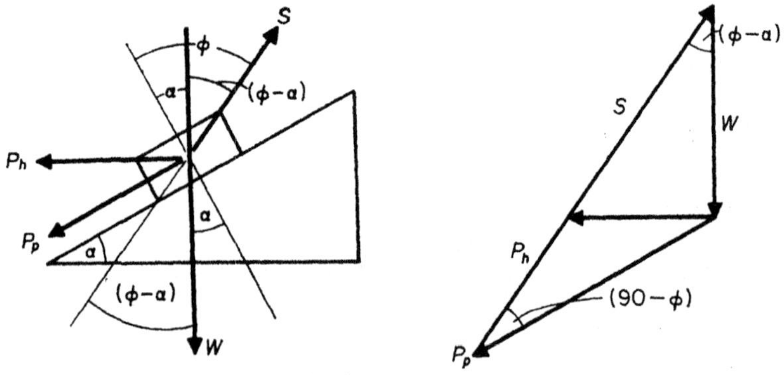

(b) Force and motion DOWN plane

(b)

Figure 6.28(*b*) Applied force tending to cause motion down the plane.

Case (*a*). The applied force, whether parallel to the plane, or horizontal, can be determined from the vector triangle. Thus,

$$P_h = W \tan(\alpha + \phi)$$

and

$$\frac{P_p}{\sin(\alpha + \phi)} = \frac{W}{\sin(90^\circ - \phi)}$$

whence

$$P_p = \frac{W \sin(\alpha + \phi)}{\cos\phi}$$

Case (*b*). Similarly, the applied force, whether parallel to the plane, or horizontal, can be determined from the vector triangle. Thus,

$$P_h = W \tan(\phi - \alpha)$$

and

$$\frac{P_p}{\sin(\phi - \alpha)} = \frac{W}{\sin(90^\circ - \phi)}$$

whence

$$P_p = \frac{W \sin(\phi - \alpha)}{\cos\phi}$$

Alternatively, the values of P_h and P_p can be measured on a vector diagram drawn to scale.

These formulae are useful, but you might prefer to insert values for the angles as you go along. This will avoid the need for complicated symbolic representations.

Example

The force required to drag a body of weight 100 lbf at a constant speed up a plane inclined at 30° to the horizontal is 68 lbf, the force being parallel to the plane. Find, graphically or otherwise, the coefficient of friction and the angle of friction. (N.C.T.E.C.)

Solution

(*a*) *Graphically*

To find the direction of S, the frictional resultant, draw the force equilibrium triangle.

The direction of S should then be marked on the space diagram, and the angle ϕ measured. Its value is 11° 45′. Then

$$\phi = 11^\circ 45', \quad \text{and} \quad \mu = \tan\phi = 0{\cdot}208$$

(*b*) *By Calculation.*

Using the sine rule,

$$\frac{100}{\sin(90 - \phi)} = \frac{68}{\sin(\alpha + \phi)}$$

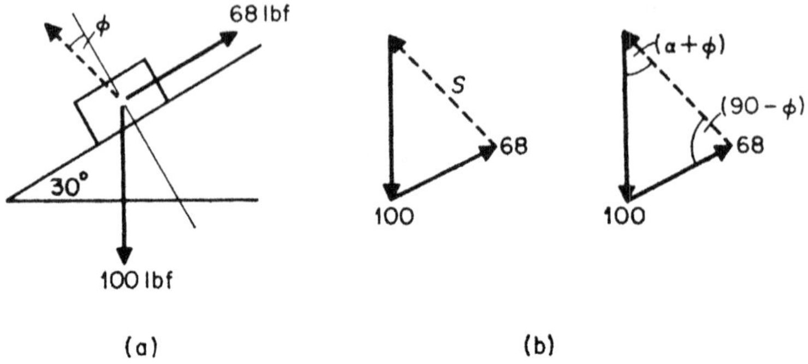

Figure 6.29(*a*) and (*b*) Space and force diagrams.

But $\alpha = 30°$

Transposing
$$68\cos\phi = 100\sin(\alpha + \phi)$$
$$= 100\sin(30° + \phi)$$

$\sin(30 + \phi)$ is a compound angle, which when expanded becomes
$$\sin(30 + \phi) = \sin 30\cos\phi + \cos 30\sin\phi$$
$$= 0{\cdot}5\cos\phi + 0{\cdot}866\sin\phi$$

$$\therefore\ 68\cos\phi = 100(0{\cdot}5\cos\phi + 0{\cdot}866\sin\phi)$$
$$= 50\cos\phi + 86{\cdot}6\sin\phi$$

That is, $18\cos\phi = 86{\cdot}6\sin\phi$

and $\tan\phi = \dfrac{18}{86{\cdot}6}$ or $0{\cdot}208$

But $\mu = \tan\phi$

$$\therefore\ \mu = 0{\cdot}208$$

and, from tables $\phi = 11°\,45'$.

Summary

Friction causes resistance to the relative motion of rubbing surfaces. It is always present wherever there is the tendency of touching bodies to slide against each other; but when sliding just occurs friction reaches a maximum value, which is called 'limiting friction'.

Laws of dry friction

1. Friction always opposes motion.
2. The force of friction depends on the nature of the surfaces in contact.

3. The magnitude of the force of friction is directly proportional to the normal reaction between the surfaces in contact.

4. The force of friction is independent of the areas of the surfaces in contact.

Coefficient of friction

This is a ratio, being the value of the force of limiting friction divided by the value of the normal reaction between the surfaces. It is denoted by the Greek μ (mu).

Angle of friction

The normal reaction and the force of friction can be combined to give a frictional resultant which we have called S in this chapter. For limiting friction the direction of this resultant is always inclined to the normal to the rubbing surfaces at an angle called the 'angle of friction' which is, for any two surfaces, constant in value. The angle of friction is denoted by the Greek symbol ϕ (phi). From the theory of this chapter, $\phi = \tan^{-1} \mu$.

Friction and the inclined plane

The resolved part of a force in a given direction is the product of the force and the cosine of the angle between the direction of the force and the given direction. Analytical solution of the problems involving friction on inclined planes can be aided by resolving all forces acting either

(*a*) horizontally and vertically, or
(*b*) parallel and perpendicular to the plane.

Angle of repose

If the inclination of a plane is increased until a block just slides on it without acceleration, then the angle made by the plane with the horizontal is called the 'angle of repose'.

This angle of inclination is equal to the angle of friction.

Geometrical and graphical solutions

By combining normal reaction and friction force to give a resultant S, most problems can be solved by drawing the equilibrium vector triangle for S, the weight W of the slider, and the applied force P. This triangle can either be scaled for an answer, or solved by the sine rule.

Exercises

1. A wooden case of weight 2 kN is pulled along a horizontal floor with uniform velocity. If the coefficient of friction between the case and the

floor is 0·3, find the force required to pull the case along when applied (*a*) horizontally (*b*) at 30° upwards from the horizontal.

2. A block of mass 100 kg rests on a plane inclined at 20° to the horizontal, and is supported partly by friction and partly by a cord parallel to the plane.

Find the pull on the cord which will (*a*) just prevent the block sliding down the plane (*b*) just produce sliding up the plane. The coefficient of friction between the two sliding surfaces may be taken as 0·3.

3. (*a*) Find the horizontal force necessary to maintain the steady motion of a block weighing 50 lbf up a plane inclined at 15° to the horizontal if the coefficient of friction between the sliding surfaces is 0·25.

(*b*) What force must be applied parallel to, and up, the plane to prevent the block from sliding down the plane.

4. A force of 1962 N is required to pull a large block of metal of mass 500 kg up a plane which is inclined at an angle of 10° to the horizontal. If the force is inclined at an angle of 20° to the plane, calculate (*a*) the coefficient of friction between the two sliding surfaces, and (*b*) the angle of friction.

5. A heavy block rests on a rough plane surface which is inclined at 50° to the horizontal, and is supported by a force P. When the force is applied parallel to the plane the block is on the point of moving UP the plane. When the force is applied horizontally the block is on the point of moving DOWN the plane. Calculate the value of the coefficient of friction between the block and the plane surface.

6. A block of mass 15 kg is placed on an inclined plane, whose height is 3 m and whose base is 4 m. If a force of 49 N is necessary, when acting in a direction parallel to the inclined plane, to prevent the block slipping down, determine the least force which, when acting in the same direction, will drag the block up the plane.

7. (*a*) In an experiment it was found that a horizontal force of 5 lbf was required to move a block weighing 15 lbf along a horizontal plane surface. Determine the coefficient of friction between the two surfaces in contact.

(*b*) The applied force was removed and the surface was then tilted until the block began to slip down the inclined plane. What was the angle between the inclined plane and the horizontal?

8. A block of stone is hauled along a horizontal wooden floor by a force inclined at an angle θ to the horizontal. If the block mass is 90 kg and the coefficient of friction between the stone and the floor is 0·3, find the effort required for values of θ varying from 0° to 90° by increments of 10 degrees. Plot a graph of force (vertical) against angle θ (horizontal), and from this determine the least value of the force, and the corresponding value of θ.

9. A block weighing 180 lbf is placed on a plane inclined at 20° to the horizontal. The block is pushed up the plane with uniform velocity by a force applied horizontally. Find the value of this force if the coefficient of friction between the block and the inclined plane is 0·3.

7

Friction and the machine

A machine is an assembly of links (*a*) for transmitting a force, (*b*) for transmitting motion, or (*c*) for doing work.

In all machines there is an input side and an output side. A force, known as the Effort, is applied to the input side to overcome a Resistance on the output side. In many machines the effect of the resistance can be likened to the raising of a load. Therefore we shall refer to the Effort applied at the input side to overcome the Load applied at the output side.

Mechanical advantage

The ratio Load/Effort is called the Mechanical Advantage. If the load and effort are determined in the same units the mechanical advantage is a number, and usually this number is greater than unity.

Velocity ratio

In order to obtain a mechanical advantage greater than unity the distance moved by the effort must be greater than the distance moved by the load.

$$\text{The ratio} = \frac{\text{Distance moved by the effort}}{\text{Distance moved by the load}}$$

is known as the velocity ratio. If these distances are determined in the same time interval the velocity ratio is a number greater than unity.

Efficiency

The efficiency of a machine is also determined as a ratio. Thus,

$$\text{Efficiency} = \frac{\text{Work taken out of the machine}}{\text{Work put into the machine}}$$

$$= \frac{\text{Output}}{\text{Input}}$$

and because,

Work done = Force × Distance moved in the direction of the force

$$\text{Efficiency} = \frac{\text{Load} \times \text{Distance moved by load}}{\text{Effort} \times \text{Distance moved by effort}}$$

$$= \frac{\text{Mechanical advantage}}{\text{Velocity ratio}}$$

Usually, efficiency is written as a percentage, and so

$$\text{Efficiency} = \frac{\text{Mechanical Advantage}}{\text{Velocity ratio}} \times 100 \text{ per cent}$$

In a perfect machine no friction has to be overcome, no moving parts have to be raised, and no slipping occurs, so

IDEAL efficiency = 100 per cent

in which case

Velocity ratio = IDEAL mechanical advantage

In practice the velocity ratio can be regarded as a constant for a particular arrangement of the machine, but the efficiency is always less than 100 per cent. Therefore the ACTUAL mechanical advantage is less than the IDEAL.

Example

A machine is used to raise a load of 120 lbf, and it is found that an effort of 24 lbf is required. At the same time as the load moves 3 in. the effort moves 18 in. Calculate (*a*) the mechanical advantage, (*b*) the velocity ratio, and (*c*) the efficiency of the machine.

Solution

$$\text{Mechanical advantage} = \frac{\text{Load}}{\text{Effort}} = \frac{120}{24} = 5$$

$$\text{Velocity ratio} = \frac{\text{Distance moved by effort}}{\text{Distance moved by load}}$$

$$= \frac{18}{3} = 6$$

$$\text{Efficiency} = \frac{\text{M.A.}}{\text{V.R.}} = \frac{5}{6} = 0{\cdot}833$$

$$= 83{\cdot}3 \text{ per cent}$$

Law of a machine

The relationship between the load at the output end and the effort applied can be found experimentally for any machine.

The results, when plotted, are found to lie on a straight line, as in figure 7.1. This line does not pass through the origin, but crosses the no-load ordinate to give a positive value for P equal to c. Further, the effort increases with load.

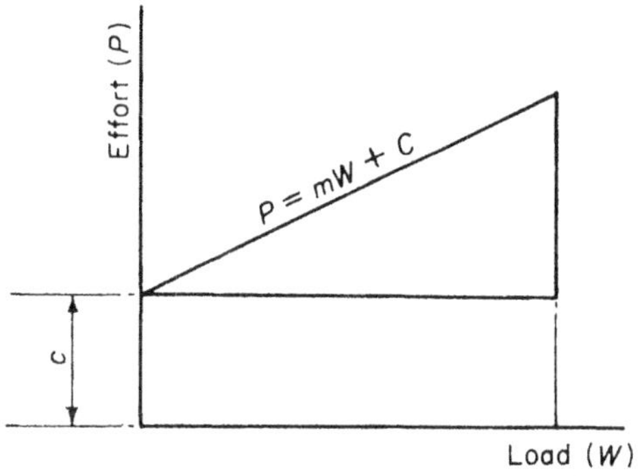

Figure 7.1 The law of a machine.

The relationship between effort and load can be represented by the equation

$$P = mW + c$$

where m is a constant defining the slope of the straight line, and c is another constant equal to the value of P where W is zero. This relationship is called 'the law of the machine'.

Note that the constant c represents an effort which has to be applied at the input end of the machine before any load can be raised at the output end, and therefore represents the force to overcome friction.

Actual mechanical advantage

The actual mechanical advantage is equal to the ratio

$$\frac{\text{Load}}{\text{Actual effort}} = \frac{W}{P}$$

If this value is plotted vertically against 'load' horizontally, we get a curve such as that shown in figure 7.2.

Limiting efficiency

Because the velocity ratio is constant for a particular arrangement, and

$$\text{Actual efficiency} = \frac{\text{Actual mechanical advantage}}{\text{Velocity ratio}}$$

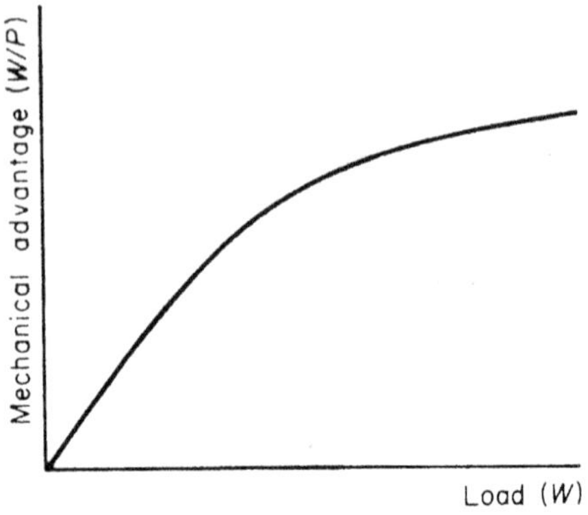

Figure 7.2 Actual mechanical advantage.

the curve obtained by plotting 'actual efficiency' against 'load' will be similar in shape to that obtained by plotting 'mechanical advantage' against 'load'.

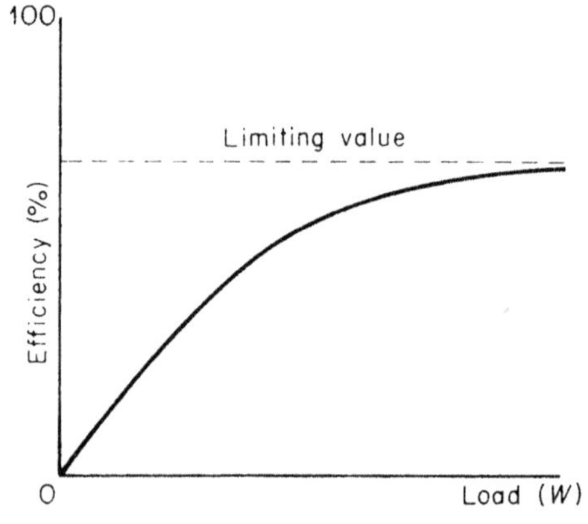

Figure 7.3 Limiting efficiency.

Note that the curve:

(a) passes through the origin,
(b) rises rapidly at first as the load increases, but
(c) later flattens out and rises very slowly with large increases in the load.

This 'flattening out' of the curve shows that the efficiency tends to reach a 'limiting value' considerably below the 100 per cent value.

In practice there is always a limiting value to the efficiency well below 100 per cent because friction is always present.

Example

The law of a small lifting machine can be represented by the equation $P = 0{\cdot}25W + 2{\cdot}5$ where $9{\cdot}8 \times P$ is the effort in N and W is the load raised in kg. If the velocity ratio of the machine is 48, calculate the effort required to raise a load of 50 kg and the distance travelled by the effort in raising the load 0·02 m.

Solution

$$\begin{aligned}\text{Effort to raise load of 50 kg} &= P \times 9{\cdot}8 \text{ N}\\ &= [(0{\cdot}25 \times 50) + 2{\cdot}5] \times 9{\cdot}81 \text{ N}\\ &= (12{\cdot}5 + 2{\cdot}5) \times 9{\cdot}8 \text{ N}\\ &= 147 \text{ N}\end{aligned}$$

$$\text{Velocity ratio} = \frac{\text{Distance moved by the effort}}{\text{Distance moved by the load}}$$

$$\begin{aligned}\text{Distance moved by the effort} &= \text{Distance moved by the load} \times \text{Velocity ratio}\\ &= 0{\cdot}02 \times 48\\ &= 0{\cdot}96 \text{ m}\end{aligned}$$

Now consider the effect of friction on some commonly used machines.

The wedge

A wedge is used mainly for forcing apart two flat surfaces. The section of a wedge can be a right angled triangle or an isosceles triangle. If a right angled triangle the wedge has a taper on one long side only. If an isosceles triangle the wedge has an equal taper on two opposite long sides.

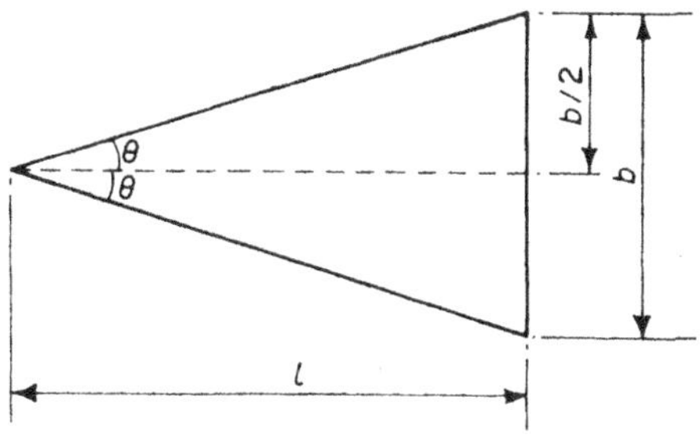

Figure 7.4 The wedge.

Consider a wedge with an equal taper on two sides, and let the angle of the wedge be 2θ.

The taper may be defined by the expression

$$\frac{b}{2} = l \tan \theta \qquad \text{(figure 7.4)}$$

or,
$$\frac{b}{l} = 2 \tan \theta \tag{7.1}$$

where b is the greatest thickness and l is the length measured perpendicular to b.

(a) Neglecting friction
For the moment, neglect friction, and consider the forces involved when the wedge is being forced between two surfaces, as shown in figure 7.5.

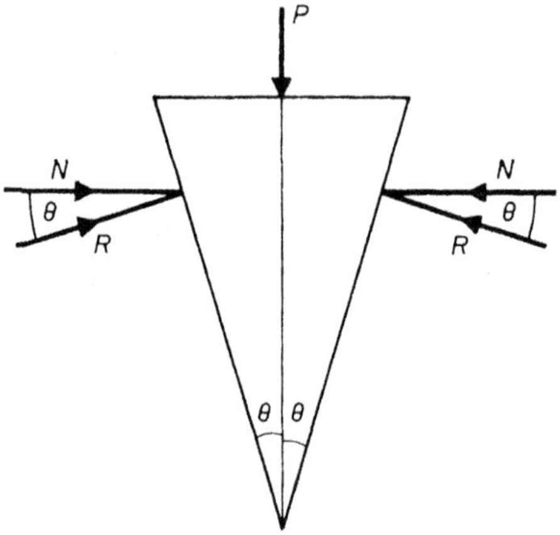

Figure 7.5 The wedge. Neglecting friction.

The driving force P is balanced by two normal reactions, each equal to R. We can determine R, if we wish, in terms of P analytically, or graphically.

Resolving the forces vertically

$$P = 2R \sin \theta \tag{7.2}$$

The true bursting, or splitting, force is N, where

$$N = R \cos \theta$$

or,
$$R = N/\cos \theta$$

Hence
$$P = \frac{2N}{\cos \theta} \times \sin \theta$$

or,
$$P = 2N \tan \theta \tag{7.3}$$

$$\text{Mechanical advantage} = \frac{\text{Splitting force}}{\text{Effort}}$$

$$= \frac{N}{P}$$

$$= \frac{1}{2 \tan \theta}$$

For 100 per cent efficiency,

$$\text{Velocity ratio} = \text{Mechanical advantage}$$

$$= \frac{1}{2 \tan \theta}$$

(*b*) *Effect of friction when driving wedge into a slit*
Consider the case when a driving force of P_d is being applied steadily to the big end of the wedge.

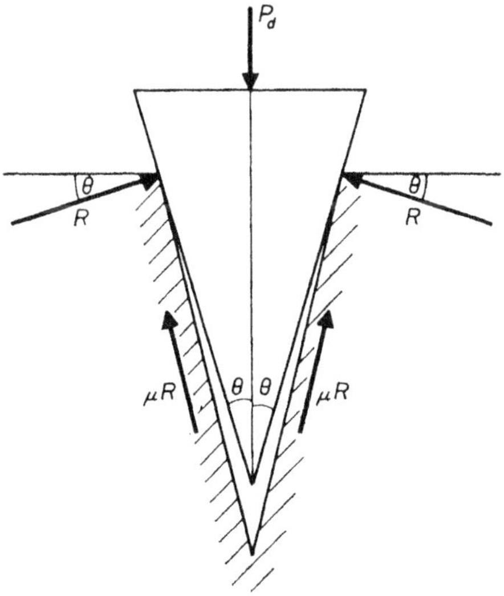

Figure 7.6 The effect of friction when driving wedge.

Let the normal reaction of the surfaces (figure 7.6) at the points of contact be R.

Then, because friction opposes motion, and the frictional force is proportional to the normal reaction, we can introduce forces equal to μR on either side of the wedge.

The forces must be in equilibrium if the force P_d is applied steadily.

If the forces are resolved vertically

$$P_d = 2[R \sin \theta + \mu R \cos \theta]$$

or,

$$P_d = 2R[\sin \theta + \mu \cos \theta]$$

which may be re-written as

$$R = \frac{P_d}{2(\sin \theta + \mu \cos \theta)} \qquad 7.4$$

(c) Effect of friction when withdrawing wedge from a slit

Suppose now that we wish to withdraw the wedge from the slit, and an axial force P_w is applied.

It is usual to assume that the value of the normal reaction R, determined in the previous section, remains the same.

Consequently the value of the frictional force opposing the withdrawal will remain as μR. Its direction, however, will be reversed as shown in figure 7.7.

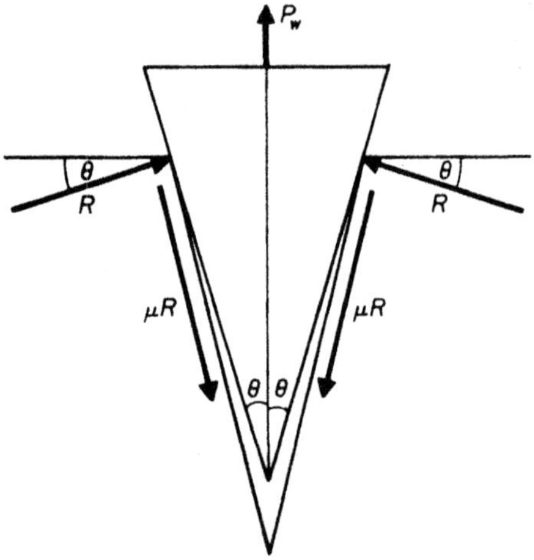

Figure 7.7 The effect of friction when withdrawing wedge.

Resolving the forces vertically, we get

$$P_w + 2R \sin \theta = 2\mu R \cos \theta$$

or,

$$P_w = 2R[\mu \cos \theta - \sin \theta]$$

and, because

$$R = \frac{P_d}{2(\sin \theta + \mu \cos \theta)} \qquad \text{from } 7.4$$

we can write

$$P_w = P_d \left(\frac{\mu \cos \theta - \sin \theta}{\sin \theta + \mu \cos \theta} \right) \qquad 7.5$$

Example

A steel wedge with a taper of 1 in 25 is used for fastening a hammer head to its shaft. If a force of 60 N is applied to the big end of the wedge, calculate the value of the force tending to split the head of the hammer. Neglect the effect of friction.

Solution

In figure 7.8 the driving force P will be resisted by the two normal reactions R.

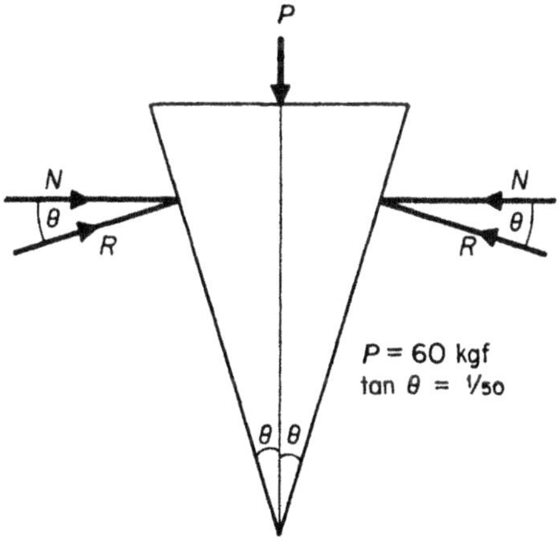

Figure 7.8 Splitting force on hammer head.

Hence, resolving vertically

$$P = 2R \sin \theta$$

The taper is defined by equation 7.1, and so

$$2 \tan \theta = \frac{b}{l} = \frac{1}{25}$$

$$\therefore \tan \theta = \frac{1}{50} = 0{\cdot}02$$

Hence $\qquad \theta = 1° \, 9'$

and $\qquad \sin \theta \doteqdot \tan \theta = 0{\cdot}02$

$$\therefore P = 2R \times 0{\cdot}02 = 60 \text{ N}$$

or, $\qquad R = \frac{60}{0{\cdot}04} = 1{,}500 \text{ N}$

The force tending to split the hammer head $= N$, and

$$N = R\cos\theta$$
$$\doteqdot R \qquad \text{because } \cos\theta \doteqdot 1$$

Hence $\qquad N = 1{,}500$ N approx.

Example

A wedge is used for forcing apart two flat surfaces. The taper of the wedge is 1 inch for each foot of its axial length. The average coefficient of friction between the wedge and the two surfaces is 0·25. A force of 500 lbf is applied to the big end of the wedge. Determine (*a*) the magnitude of the forces normal to the wedge surfaces resisting separation, and (*b*) the magnitude of the force necessary to withdraw the wedge before the two surfaces have separated.

Solution

(*a*) Let figure 7.9 represent the forces acting on the wedge as it is being driven between the two surfaces.

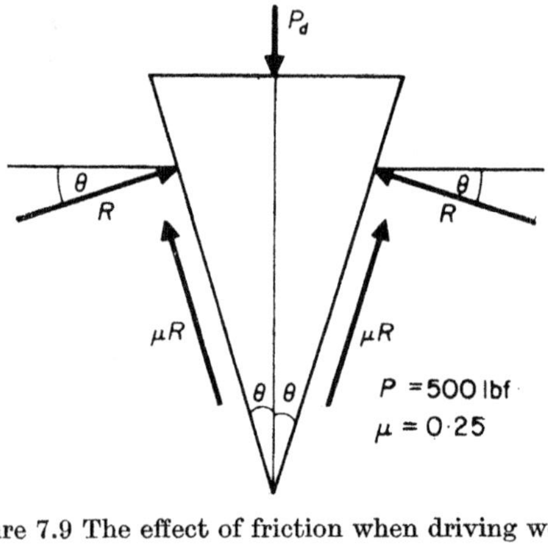

Figure 7.9 The effect of friction when driving wedge.

Because the taper of the wedge is 1 in 12, we can write

$$2\tan\theta = \frac{1}{12}$$

or, $\qquad \tan\theta = \frac{1}{24}$

or, $\qquad \theta = 2^\circ\ 24'$

whence $\qquad \sin\theta = 0{\cdot}042$

and $\qquad \cos\theta = 0{\cdot}999$

If we resolve forces vertically

$$P_d = 2\mu R \cos\theta + 2R \sin\theta$$
$$= 2R(\mu \cos\theta + \sin\theta)$$

or, $$R = \frac{P_d}{2(\mu \cos\theta + \sin\theta)} \quad \text{as in equation 7.4}$$
$$= \frac{500}{2(0{\cdot}25 + 0{\cdot}042)}$$
$$= \frac{250}{0{\cdot}292}$$
$$= 857 \text{ lbf}$$

(*b*) Using equation 7.5, or working from first principles,

$$P_w = 2R(\mu \cos\theta - \sin\theta)$$
$$= 1{,}714\ (0{\cdot}25 - 0{\cdot}042)$$
$$= 1{,}714 \times 0{\cdot}208$$
$$= 357 \text{ lbf}$$

Some simple lifting mechanisms—Pulleys

The simplest type of pulley block is the single pulley, as shown in figure 7.10. Here the load W applies a tension to a cord or a rope,

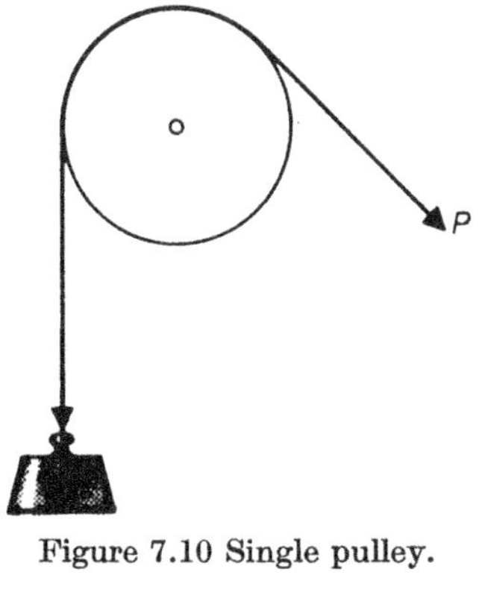

Figure 7.10 Single pulley.

which, if there were no frictional losses, would be the same as the effort P at the hands. The greatest load which a man could lift under these circumstances would be equal to his own weight.

From examination of the diagram it will be apparent that the Velocity Ratio of the single pulley is 1.

Pulleys can be compounded to give a higher velocity ratio, as shown in figure 7.11, where pulley A is suspended in a loop of the cord, which then passes over pulley B.

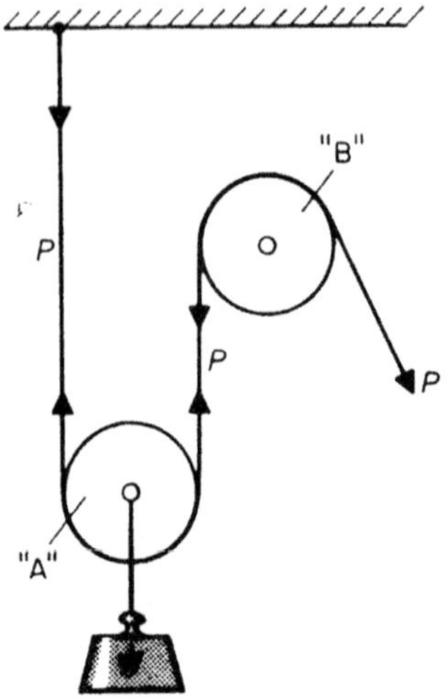

Figure 7.11 Compound pulley.

Effort P is applied at the free end of the cord, and load W suspended from pulley A. If no friction is present (that is, assuming 100 per cent efficiency) then, considering the equilibrium of the bottom pulley, it can be seen that

$$2P = W$$

or, Mechanical advantage = 2

Since the efficiency has been taken to be 100 per cent then

Mechanical advantage = Velocity ratio = 2

a fact which can be verified experimentally.

It is usual to find more than one pulley in a compound sheave. In figure 7.12 the cord is continuous.

In all such cases it can be shown, by considering the equilibrium of the bottom, or movable, sheave, the velocity ratio is equal to the number of suspending cords of the bottom sheave. In the diagram this is 4. But this number must be twice the number of pulleys in the bottom sheave. Thus, FOR ALL PULLEY BLOCKS OF THE COMPOUND TYPE WHERE THE CORD IS CONTINUOUS.

Velocity ratio = 2 × Number of pulleys in bottom sheave

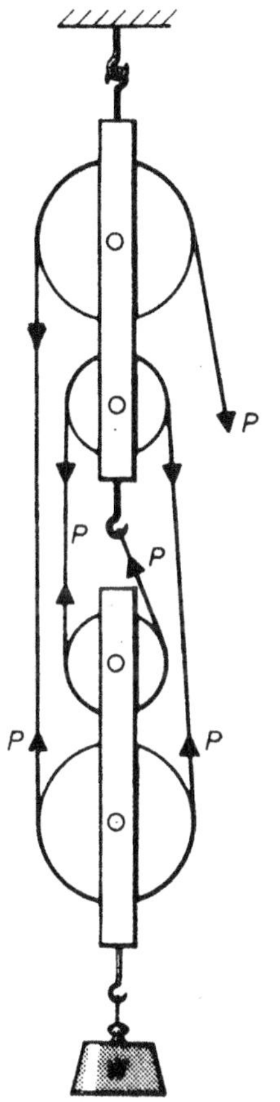

Figure 7.12 Compound sheaves—fixed and movable.

Example

A compound pulley with a continuous cord has three pulleys per sheave. What is its Velocity Ratio?

Solution

$$\begin{aligned} \text{Velocity ratio} &= 2 \times \text{Number of pulleys in bottom sheave} \\ &= 2 \times 3 \\ &= 6 \end{aligned}$$

The next paragraph shows the effect where the cord is NOT continuous.

A further pulley system

Suppose we want to raise a load W by an effort P applied to a rope at one end of the pulley system shown in figure 7.13.

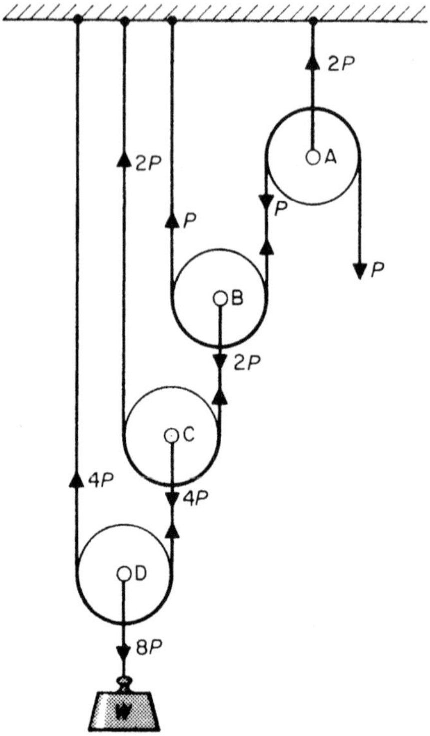

Figure 7.13 Multi-pulley system with no friction.

Consider first of all a system without friction.

(a) No friction

If there is no friction at the pulley bearings, and no slipping of the ropes over the pulleys, the forces in the rope on either side of a pulley must be equal, and the force in the suspension rope at the centre of each pulley must be double the force which passes round the pulley.

Therefore, the force in each of the ropes can be determined if we start from the end at which the effort P is applied.

Because the forces in the two sides of the rope over pulley A are each equal to P, the force in the suspension rope for this pulley must be $2P$. Similarly, because the forces in the two sides of the rope over pulley B are each equal to P, the force in the rope from the centre of this pulley and passing round pulley C must be equal to $2P$. Likewise, the force in the rope from the centre of pulley C and passing round pulley D must be equal to $4P$, and consequently the force in the rope carrying the load W must be equal to $8P$. That is,

$$W = 8P$$

or, $$\text{Mechanical advantage} = \frac{\text{Load}}{\text{Effort}} = \frac{W}{P} = 8$$

If the load W is raised by 1 unit of length, the pulley D must be raised by the same amount. Because the lengths of the ropes are unchanged, pulley C must consequently be raised by 2 units of length, pulley B by 4 units and the free end of the rope round pulley A by 8 units. That is,

$$\text{Velocity ratio} = \frac{\text{Distance moved by effort}}{\text{Distance moved by load}}$$
$$= 8$$

and $$\text{Efficiency} = \frac{\text{Mechanical advantage}}{\text{Velocity ratio}} \times 100 \text{ per cent}$$
$$= 100 \text{ per cent}$$

(*b*) *With friction*

Suppose that in the pulley system shown in figure 7.14 a force F is required to overcome friction in each of the pulley bearings, and hence to turn each pulley.

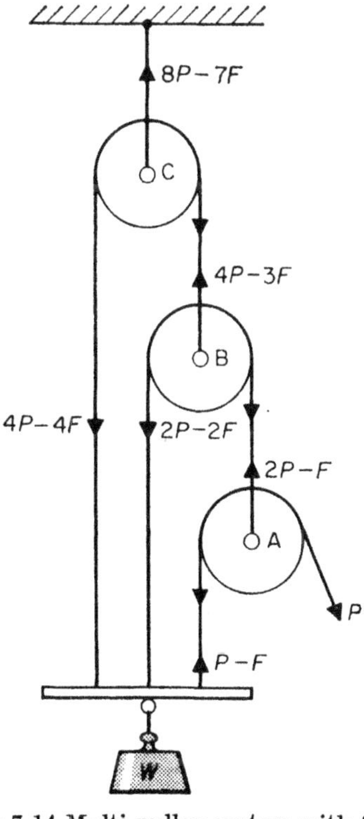

Figure 7.14 Multi-pulley system with friction.

We may, as in (*a*) determine the forces in the ropes on either side of each pulley by starting at the free end of the rope where the effort P is applied.

Because of friction the force in the rope on the left hand side of pulley A must be $(P - F)$ and so the total force in the rope by which pulley A is suspended must be

$$P + (P - F) = 2P - F$$

Similarly the force in the rope on the left hand side of pulley B must be $(2P - 2F)$, and so the total force in the rope by which pulley B is suspended must be

$$(2P - F) + (2P - 2F) = 4P - 3F$$

Similarly the force in the rope on the left hand side of pulley C must be $(4P - 4F)$, and so the total force in the rope by which pulley C is suspended must be

$$(4P - 3F) + (4P - 4F) = 8P - 7F$$

The upward force on the plate which carries the load W, is, therefore,

$$\begin{aligned}(4P - 4F) + (2P - 2F) + (P - F) &= 7P - 7F \\ &= 7(P - F)\end{aligned}$$

and this must be equal to the load W. Hence

$$7(P - F) = W$$

or,

$$P = \frac{W + 7F}{7}$$

$$\begin{aligned}\text{Mechanical advantage} &= \frac{W}{P} \\ &= \frac{7W}{W + 7F}\end{aligned}$$

The velocity ratio is unaffected by friction, and so may be found conveniently by considering the effort required when the system is perfect—that is, when no friction is present.

Thus, for an efficiency of 100 per cent, the forces in the ropes on the left hand sides of the pulleys would be P, $2P$ and $4P$, and so the upward force on the plate would be

$$P + 2P + 4P = 7P$$

and this must be equal to the load W. Hence

$$7P = W$$

and

$$\begin{aligned}\text{Mechanical advantage of the perfect system} &= \frac{W}{P} \\ &= 7\end{aligned}$$

In a perfect system, the mechanical advantage is equal to the velocity ratio, and so

$$\text{Velocity ratio of the system} = 7$$

When friction is present the velocity ratio is unchanged. Therefore,

$$\text{Efficiency of pulley system with friction} = \frac{\text{M.A.}}{\text{V.R.}}$$

$$= \frac{7W}{7(W + 7F)}$$

$$= \frac{W}{W + 7F}$$

or,

$$\left(\frac{W}{W + 7F}\right) \times 100 \text{ per cent}$$

Differential pulleys—The differential wheel and axle. Weston's block

A useful design of pulley, much used for heavy lifting, is the differential pulley, typically found in the Weston's block. An experimental investigation to demonstrate the principle is available in the differential

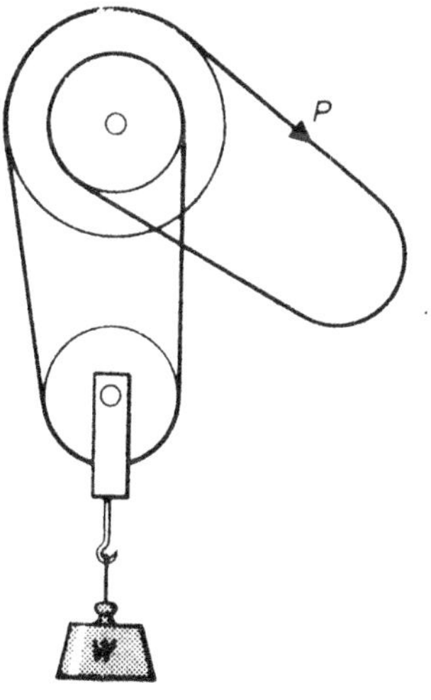

Figure 7.15 Weston's pulley block.

wheel and axle. (See page 65 of the book General Engineering Science in this series.)

In this type (figure 7.15) an endless cord is wrapped round a two-diameter pulley in the top sheave, and carries in its lower loop the bottom (single pulley) sheave.

The effect of the arrangement is that, as the cord causes the top two-diameter pulley to rotate, it tends to shorten the hanging loop one periphery of the large top wheel for each rotation of that wheel. But the loop is meanwhile allowed to lengthen by an amount equal to the smaller periphery of the top wheel. This results in a total shortening of the loop cord by an amount equal to the difference of the large and small peripheries of the top sheave. This shortening leads in turn to a raising of the hanging sheave by half the amount of the shortening. The Velocity ratio of the system is therefore

$$\frac{\text{Large periphery}}{\text{Half the difference of the peripheries}}$$

If the two diameters of the top sheave are D and d, then

$$\text{Velocity ratio} = \frac{\pi D}{\pi(D-d)} \times 2 = \frac{2D}{(D-d)} \qquad 7.6$$

Example

A Weston's block has a differential wheel of diameters 10 in. and 9 in. What is its Velocity Ratio?

Solution

$$\text{Velocity ratio} = \frac{2 \times 10}{(10-9)} = 20$$

Example

A geared Weston's block has a differential wheel of 25 teeth and 23 teeth. What is its Velocity ratio?

Solution

$$\text{Velocity ratio} = \frac{2 \times 25}{(25-23)} = 25$$

Consideration of the efficiency of a Weston's pulley block in its conventional form reveals a nice principle of design which could be used in all lifting machinery.

IF THE BLOCK IS MADE SO THAT IT FUNCTIONS WITH LESS THAN 50 PER CENT EFFICIENCY, NO PRECAUTIONS NEED BE TAKEN AGAINST ACCIDENTAL RUN-DOWN OF THE LOAD.

This run-down, called 'overhauling' is dealt with in the following section.

Overhauling

If, when a load is being raised, the effort is removed from the input end of the machine, the load may then overcome any existing frictional forces and begin to fall, thus reversing the original effect of the effort. Such a machine is said to overhaul.

We have seen that in the ideal frictionless machine,

$$\text{Velocity ratio} = \text{Ideal mechanical advantage}$$

$$= \frac{\text{Load}}{\text{Ideal effort}}$$

Hence,
$$\text{Ideal effort} = \frac{\text{Load}}{\text{Velocity ratio}}$$

In an actual machine, friction—which always opposes the motion—makes it necessary to exert a greater effort and so, because the velocity ratio is constant, friction can be regarded as an extra load.

$$\therefore \text{ Actual effort} = \text{Ideal effort to overcome load} + \text{Effort to overcome friction}$$

or,

$$\therefore \text{ Actual effort} = \frac{\text{Load being raised} + \text{Extra load due to friction}}{\text{Velocity ratio}}$$

$$P_{\text{actual}} = \frac{W + W_f}{\text{V.R.}} \qquad \text{(see figure 7.16)}$$

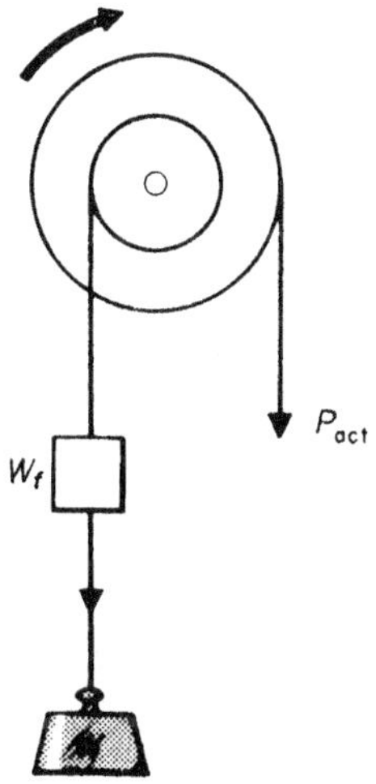

Figure 7.16 Effort applied—load rising.

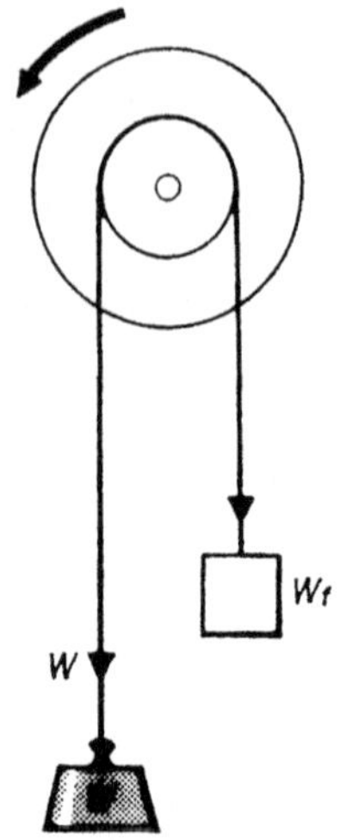

Figure 7.17 No effort applied—load falling.

If now the effort is removed, the load will tend to fall, as in figure 7.17, and, because friction opposes motion it will resist the falling of the load.

If W_f is greater than W the load will not fall. If W_f is equal to, or less than W, the load will fall, or 'overhaul'. Hence the condition for overhauling is that the frictional force must be equal to, or less than, the load being raised.

Suppose that $W_f = xW$

Then
$$P_{\text{actual}} = \frac{W + xW}{\text{V.R.}}$$
$$= \frac{W}{\text{V.R.}}(1 + x)$$
$$\therefore\ P_{\text{act}} \times \text{V.R.} = W(1 + x) \qquad 7.7$$

But,
$$\text{Actual efficiency} = \frac{\text{Act. mech. advantage}}{\text{Velocity ratio}} \times 100 \text{ per cent}$$
$$= \frac{\text{Load raised}}{\text{Actual effort}} \times \frac{100}{\text{V.R.}} \text{ per cent}$$
$$\eta_{\text{act}} = \frac{W \times 100}{P_{\text{act}} \times \text{V.R.}} \text{ per cent}$$
$$= \frac{W \times 100}{W(1 + x)} \text{ per cent, from 7.7}$$
$$= \frac{100}{1 + x} \text{ per cent}$$

If $x < 1$ (i.e. friction force < load being raised)

$$\therefore \; \eta_{actual} > 50 \text{ per cent}$$

If $x = 1$ (i.e. friction force = load being raised)

$$\eta_{actual} = 50 \text{ per cent}$$

If $x > 1$ (i.e. friction force > load being raised)

$$\eta_{actual} < 50 \text{ per cent}$$

Consequently, if the load is NOT to overhaul, the efficiency of the machine must be less than 50 per cent.

(*Note:* The low efficiencies associated with the screw type of mechanism make it unnecessary to guard against rundown of the load.)

Winches

The winch is a common mechanism used for winding and hauling by means of a cable.

The effort is applied to the end of a handle which provides the leverage, and the cable which hauls the load is wound on to a drum (see figure 7.18).

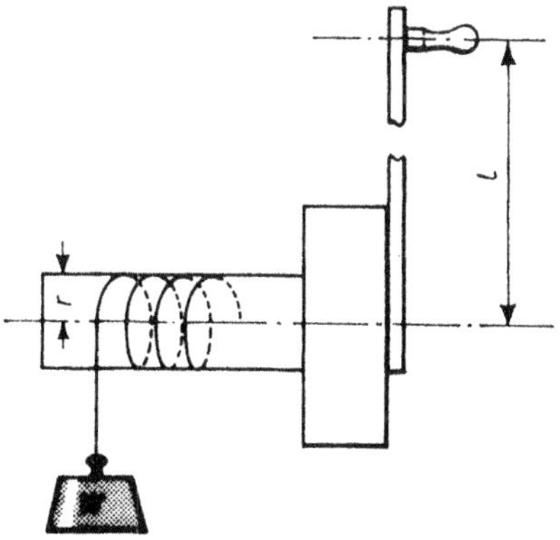

Figure 7.18 The winch.

In ungeared winches the velocity ratio can be seen to be the direct ratio of the handle length to the drum radius. Thus, for each rotation of the handle

$$\frac{\text{Distance moved by effort}}{\text{Distance moved by load}} = \frac{2\pi l}{2\pi r} = \frac{l}{r}$$

In geared winches, simple or compound gears are used to multiply the above velocity ratio, and increase mechanical advantage.

Example

What is meant by overhauling when applied to a lifting machine? State the conditions which must exist in order that overhauling may take place.

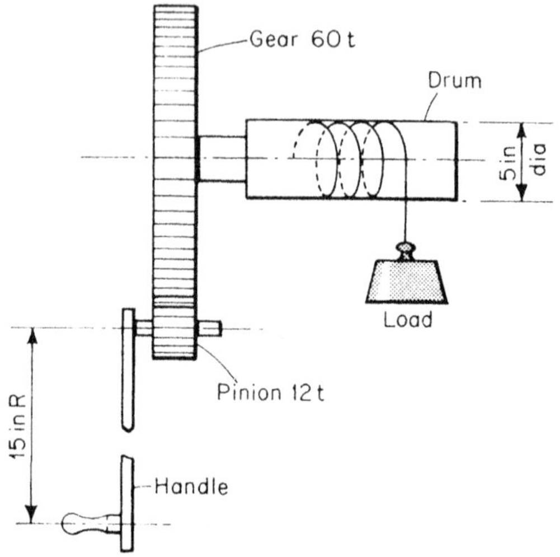

Figure 7.19 The crab winch.

An experimental crab winch is shown in figure 7.19. The following results were obtained in a test on the winch.

Load (lbf)	20	30	40	50	60
Effort (lbf)	1·7	2·4	2·5	2·9	3·3

Plot a graph to show how the efficiency of the winch varies with the load lifted.

What is the maximum load that the winch can take if overhauling is to be avoided? (U.L.C.I.)

Solution

It will be convenient to tabulate the calculation, including the efficiency figures.

Consider one revolution of the handle.

$$\text{Velocity ratio} = \frac{\text{Distance moved by effort}}{\text{Distance moved by load}}$$

$$= \frac{2\pi \times 15}{2\pi \times \frac{5}{2} \times \frac{12}{60}} = 30$$

Tabulation

Load lbf	20	30	40	50	60
Effort lbf	1·7	2·4	2·5	2·9	3·3
M.A. = Load/Effort	11·7	12·5	16	17·2	18·2
V.R.	30	30	30	30	30
Efficiency per cent $\eta = \frac{\text{M.A.}}{\text{V.R.}} \times 100$	39	42	53	57	61

It will be seen from the table that the exact load at which overhauling will occur is between 30 lbf and 40 lbf. It can be obtained more accurately from a graph (figure 7.20).

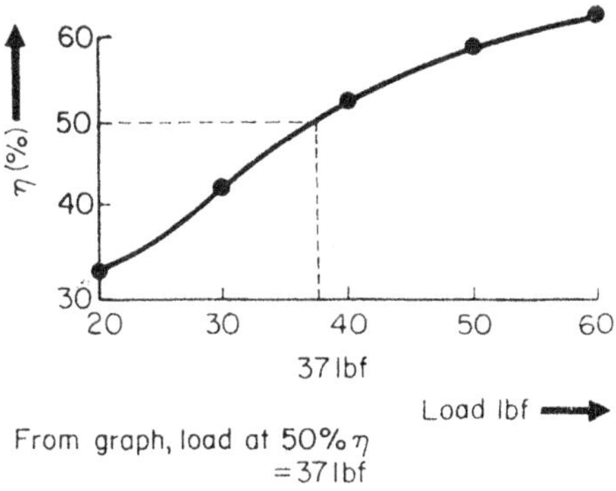

Figure 7.20 Efficiency—load graph.

From the graph, load at 50 per cent efficiency is 37 lbf. This is the load which will overhaul.

Screw threads

If a straight line, inclined to the horizontal, is drawn on a piece of paper and the paper is then wrapped round the outside of a cylinder, the line takes the form of a helix. The profile of a screw thread is a helix, so that if one complete turn of the helix could be unwrapped it would appear as an inclined line AB, where AC is drawn at right angles to the axis of the bolt, or screw.

If AC represents the mean circumference of the thread, then BC must represent the lead of the screw—the lead being the distance moved axially in one revolution.

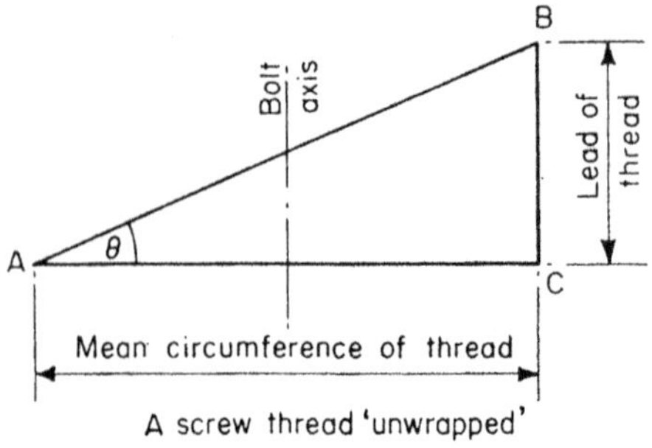

Figure 7.21 A screw thread 'unwrapped'.

Tightening the nut on a bolt

Suppose the helix angle of the thread of a bolt is θ, and that a nut is being tightened up under an axial load W. The conditions are equivalent to pushing a weight W up a plane inclined at the same angle to the horizontal, and shown diagrammatically in figure 7.22.

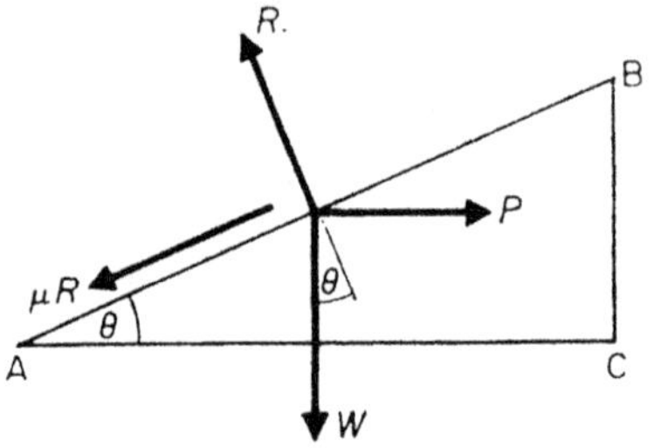

Figure 7.22 Tightening a nut.

A force P, tangential to the mean radius of the bolt is applied in a horizontal direction by means of a lever. The frictional force is μ times the normal reaction and opposes motion up the plane.

We can obtain a relationship between W and P by resolving the forces in two directions at right angles and eliminating R.

Resolving perpendicular to the inclined plane

$$R = P \sin \theta + W \cos \theta \tag{7.8}$$

Resolving parallel to the plane

$$\mu R + W \sin \theta = P \cos \theta \tag{7.9}$$

Substituting the value of R from 7.8 into 7.9, we get

$$\mu P \sin \theta + \mu W \cos \theta + W \sin \theta = P \cos \theta$$

or,

$$W(\mu \cos \theta + \sin \theta) = P(\cos \theta - \mu \sin \theta)$$

That is,

$$\text{Mechanical advantage } \frac{W}{P} = \frac{\cos\theta - \mu\sin\theta}{\mu\cos\theta + \sin\theta}$$

$$\text{Work done by effort} = \text{Effort} \times \text{Distance moved}$$
$$= P \times \text{AC}$$

and

$$\text{Work done by load} = W \times \text{BC}$$

But,

$$\text{Efficiency} = \frac{\text{Work done by load}}{\text{Work done by effort}}$$
$$= \frac{W}{P} \times \frac{\text{BC}}{\text{AC}}$$
$$= \frac{W}{P} \times \tan\theta$$

Note that

$$\frac{\text{AC}}{\text{BC}} = \text{Velocity ratio}$$

and so

$$\text{Velocity ratio} = \frac{1}{\tan\theta} = \frac{\text{Mean circumference of thread}}{\text{Lead}}$$

Screw jack

A screw jack is just another form of tightening a nut on a bolt. The effort is here applied at the end of a lever arm, and

$$\text{Velocity ratio} = \frac{\text{Distance moved by effort}}{\text{Distance moved by load}}$$

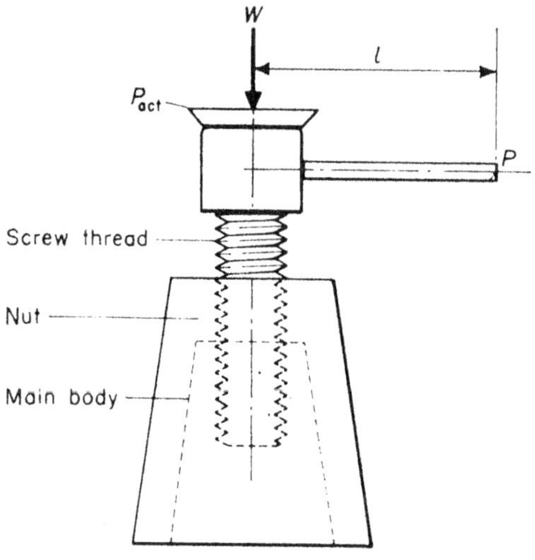

Figure 7.23 Screw jack.

In one revolution, the effort P (figure 7.23) moves a distance $2\pi l$, where l is the distance of the point of application of P from the axis of the screw, whereas the load W moves a distance equal to the lead of the screw thread in an axial direction.

Note particularly that, in a screw jack, the load must not fall when the effort is removed, and so, referring to the condition for the prevention of overhauling, the efficiency of the screw jack must be less than 50 per cent.

If the load W does not rotate with the nut, the latter must rub against a collar or pad, and so there will be friction between the nut and the pad. In a screw jack the pad carrying the load is often stationary.

Suppose that

r_n = Mean radius of the nut,

and μ_n = Coefficient of friction between the nut face and the pad carrying the load W.

Then,

$$\text{Frictional force at the nut face} = \mu_n W$$
$$\text{Torque necessary to overcome this force} = \mu_n W r_n$$

Example

The design of a screw jack is such that an ideal effort of 15 lbf should be required at a distance of $13\frac{3}{8}$ in. from the axis of the screw to lift a load of 15 cwt. In practice it is found that an effort of 37·5 lbf is required. Calculate the efficiency of the jack under these conditions.

Solution

$$\text{Velocity ratio} = \frac{\text{Distance moved by effort}}{\text{Distance moved by load}}$$
$$= \frac{2\pi \times 13\frac{3}{8}}{\text{Lead (in.)}}$$
$$= \frac{2 \times \frac{22}{7} \times \frac{107}{8}}{l}$$

where l = Lead of screw (in.)

If the efficiency is 100 per cent

$$\text{V.R.} = \text{M.A.} = \frac{\text{Load}}{\text{Ideal effort}}$$
$$= \frac{15 \times 112}{15}$$
$$= 112$$

That is,

$$\frac{2 \times \frac{22}{7} \times \frac{107}{8}}{l} = 112$$

or, $$l = 2 \times \frac{22}{7} \times \frac{107}{8} \times \frac{1}{112}$$

$$= 0{\cdot}75 \text{ in.}$$

But $$\text{Actual mechanical advantage} = \frac{15 \times 112}{37{\cdot}5}$$

$$\text{Actual efficiency} = \frac{\text{Actual M.A.}}{\text{V.R.}} \times 100 \text{ per cent}$$

$$= \frac{15 \times 112}{37{\cdot}5 \times 112} \times 100 \text{ per cent}$$

$$= 40 \text{ per cent}$$

Reducing the effect of friction

The Conservation of Energy Law states that energy cannot be created or destroyed, but can only change its form. A change of energy is equal to the work done. Therefore, the energy absorbed, or the work done in overcoming friction must appear in some other form. In a friction brake or in a bearing it appears as heat energy. This heat energy is wasted energy and decreases the efficiency of the machine. Therefore, if we are to maintain a high efficiency we must aim at reducing the value of the coefficient of friction.

In plain bearings this is achieved by the use of a soft metal lining which has a low coefficient of friction; the lining, of course, being easy to replace when worn out.

In ball bearings the surfaces of the balls are made very hard so that they do not distort, and their resistance to rolling is very small. Note that the coefficient of rolling friction is much less than the coefficient of sliding friction.

Lubrication

Another method of reducing the loss of energy in overcoming friction is by the use of a lubricant.

In the previous chapter it was stated that all surfaces are rough, and friction occurs when one such dry surface is rubbed over another.

If sufficient lubricant is applied to the surfaces a film is formed separating the two rough surfaces. Consequently, if one surface is now rubbed over the other, and the film is maintained, the resistance to motion will depend entirely on the lubricant, and not on the nature of the two surfaces.

In a bearing the aim should be to separate the rotating shaft, or journal, from the bearing surfaces by a film of lubricant. If the load becomes excessive the film is liable to break down, and contact would then take place between the two metals concerned. High temperatures would occur locally at high spots on the rough surfaces, and this would lead to melting of the metal, and possibly to seizure.

Work done by a force. Power

Work is done when a force moves through a distance. If, therefore, the load is raised by a given distance, or at a given speed, work is done, and the amount of work done on the load is equal to the weight raised multiplied by the distance moved. In order to raise this weight work is done by the effort at the input end of the machine.

In an ideal machine with an efficiency of 100 per cent, the work done by the effort is equal to the work done on the load. However, because of friction, the amount of work put into the machine is more than the amount of work taken out. Therefore, if we are asked to determine the work done in raising the load, we must evaluate the WORK DONE BY THE EFFORT. Thus,

$$\text{Work done by effort} = \text{Effort} \times \text{Distance moved by the effort}$$

and, because

$$\text{Velocity ratio} = \frac{\text{Distance moved by effort}}{\text{Distance moved by load}}$$

$$\text{Work done by effort} = \text{Effort} \times \text{Distance moved by LOAD} \times \text{Velocity ratio}$$

Power is the rate of doing work. Consequently,

$$\text{Power required at input end} = \text{Effort} \times \text{Speed at which LOAD rises} \times \text{Velocity ratio}$$

Power units: the horsepower and the watt

If the effort is determined in lbf and the speed in ft/min, the power is obtained in ft lbf/min.

But 33,000 ft lbf/min are equivalent to 1 horsepower.

$$\therefore \text{Horsepower} = \frac{\text{Effort (lbf)} \times \text{Speed at which load rises (ft/min)} \times \text{V.R.}}{33{,}000}$$

On the SI system, the work unit is the metre-newton, or joule: and the power unit is the watt, a rate of working of one joule per second.

Work done by a torque

In many machines work is transmitted by rotation of a shaft, or a pulley. If we are to transmit power by rotation a force must be applied to the end of a radius arm at the input end so that a turning moment is given to the shaft. Torque is another name for a turning moment, and its value is equal to the force producing rotation multiplied by the length of the radius arm. Thus, in figure 7.24:

$$\begin{aligned}\text{Torque, or turning moment} &= T \\ &= Pr\end{aligned}$$

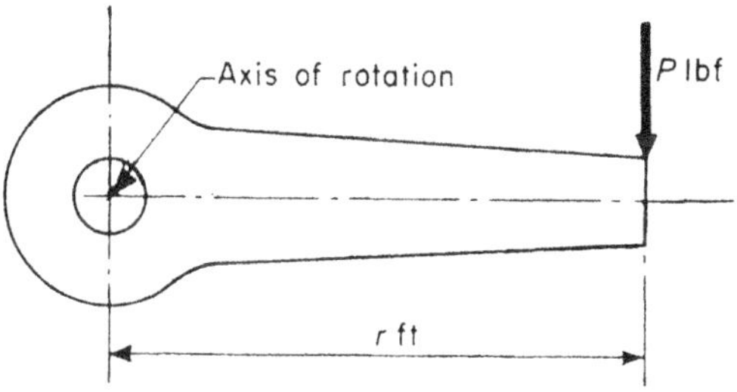

Figure 7.24 A torque.

If the effort is determined in lbf and the radius in ft the torque will be obtained in lbf ft.

Note the units 'lbf ft' and distinguish them from 'ft lbf' which are used for work done.

Work is done by a torque when it turns through an angle, its value being equal to the product of the torque (T) and the angle (θ) turned through in radians. Thus,

$$\begin{aligned}\text{Work done by torque} &= T\theta \\ &= Pr\theta \text{ ft lbf}\end{aligned}$$

If the radius arm is being rotated at n_m revolutions per minute,

$$\theta = 2\pi n_m \text{ radians per minute}$$

$$\therefore \text{ Work done by torque in 1 minute} = Pr \times 2\pi n_m \text{ ft lbf/min}$$

and, because 33,000 ft lbf/min are equivalent to 1 horsepower,

$$\begin{aligned}\text{Horsepower} &= \frac{Pr \times 2\pi n_m}{33{,}000} \\ &= \frac{2\pi n_m T}{33{,}000}\end{aligned}$$

Example

A shaft transmits 20 horsepower at 2,000 rev/min. Calculate the torque on the shaft.

Solution

$$\text{Horsepower} = \frac{2\pi n_m T}{33{,}000}$$

$$\therefore \text{ Torque} = \frac{33{,}000 \times \text{hp}}{2\pi n_m} \text{ lbf ft}$$

$$= \frac{33{,}000 \times 7 \times 20}{2 \times 22 \times 2000}$$

$$= 52{\cdot}5 \text{ lbf ft}$$

Friction in a plain bearing

We have seen, in Chapter 6, that when a block of weight W lbf is caused to move over a rough horizontal surface by the action of an external force P lbf the limiting value of the frictional force opposing the motion is equal to μW lbf, where μ is the coefficient of friction.

A similar condition applies when a shaft rotates in a cylindrical bearing. Thus, suppose a shaft is rotating in a clockwise direction in a plain bearing, as in figure 7.25, where the clearance is greatly exaggerated.

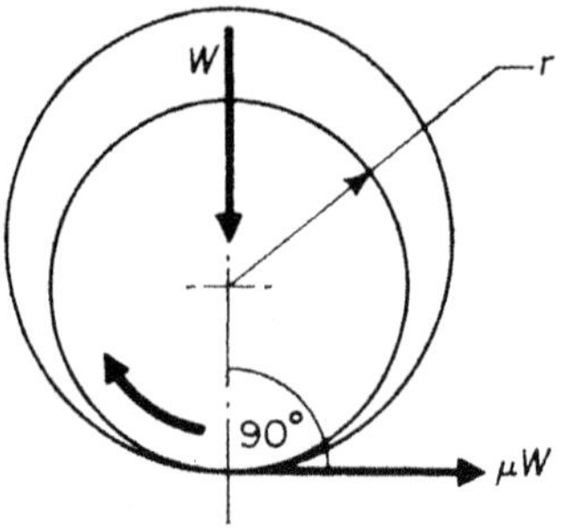

Figure 7.25 A plain bearing.

If W is the vertical load on the shaft, a frictional force equal to μW acts at right angles to W and opposes the motion. This frictional force is tangential to the shaft, and, because it acts at a distance r from the centre of the shaft, will produce a frictional torque opposing the rotation. This frictional torque must be overcome by the effort applied to the shaft before it can do useful work, and so a small fraction of the horsepower transmitted by the shaft will be used up in overcoming this torque.

$$\text{Frictional torque} = \mu W \times r \text{ lbf ft}$$

where r = radius of shaft in feet

If the shaft is rotating at n_m rev/min,

$$\text{Work done by this torque} = \mu W r \times 2\pi n_m \text{ ft lbf/min}$$

Horsepower absorbed in overcoming friction

$$= \frac{\mu W \times 2\pi r n_m}{33{,}000}$$

Note that the value of μ for a smooth cylindrical bearing is low, and usually lies between 0·02 and 0·03. Contrast this value with that for two rough plane surfaces sliding over each other, where the value of μ lies between 0·25 and 0·5—that is, more than 10 times the value of μ for a bearing.

Transmission of power by belts

Suppose that we have two pulleys connected by an endless belt as in figure 7.26.

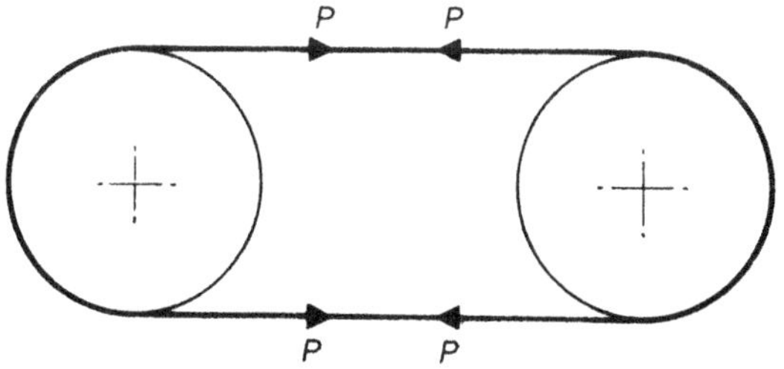

Figure 7.26 Endless belt. Stationary pulleys.

If the two pulleys are stationary the tension in the upper portion of the belt is the same as that in the lower portion. The actual value of the tension (P) can be adjusted by changing the distance between the centres of the two pulleys.

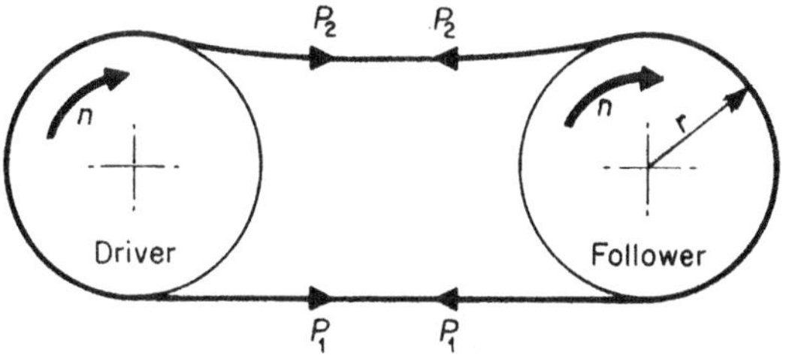

Figure 7.27 Endless Belt transmitting power.

If, however, the pulleys are rotating, torque is applied to the driver, as in figure 7.27, causing the portion of the belt coming on to the driver to become tight, and the portion of the belt leaving the driver to become slack.

The difference in the tensions is appreciable, that on the tight side being almost three times that on the slack side.

This difference in tensions is only made possible by the frictional forces between the pulley and the belt. If there were no friction the driving pulley would slip and no power would be transmitted by the belt.

Consider what happens to the follower when power is being transmitted.

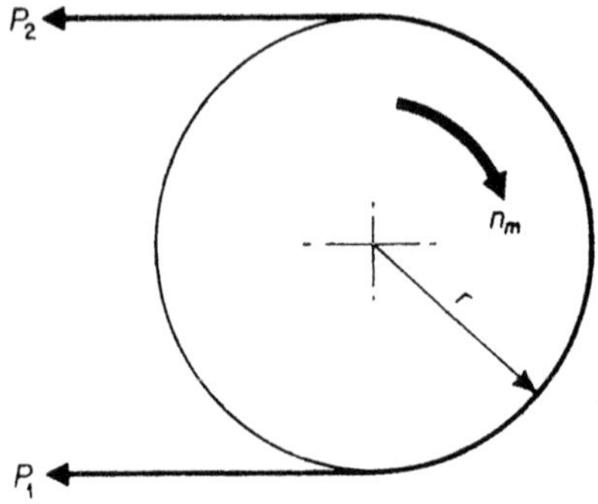

Figure 7.28 The follower.

Let P_1 = Tension in belt on the tight side (lbf)
P_2 = Tension in belt on the slack side (lbf)
r = Radius of follower (ft).

$$\text{Clockwise torque on follower} = P_1 r \text{ lbf ft}$$
$$\text{Anti-clockwise torque on follower} = P_2 r \text{ lbf ft}$$
$$\text{Resultant driving torque} = (P_1 - P_2) r \text{ lbf ft}$$

If the follower is being rotated clockwise at n revolutions per minute.

$$\text{Horsepower transmitted by belt} = (P_1 - P_2) r \times \frac{2\pi n_m}{33{,}000}$$

Note that the linear speed of the circumference of the follower (and hence of the belt if no slip occurs) is $2\pi n_m r$ ft/min, and that

$$\text{Horsepower transmitted} = \frac{\text{Effective force (lbf)} \times \text{Linear speed (ft/min)}}{33{,}000}$$

$$\text{Power in watts} = \text{Effective force } (N) \times \text{Linear speed (m/s)}$$

Example

Show by means of a sketch how a belt transmits motion to a pulley and state clearly what is meant by "effective pull" in terms of the tensions of the belt.

A belt drive is required to transmit 30 hp from a shaft running at 100 rev/min to one running at 250 rev/min. Find the tensions in the tight and slack sides of the belt, assuming the smaller pulley to be 24 in. diameter and the ratio of the tensions to be 2·5 to 1.

Determine also a suitable width of belt if the allowable working tension is 120 lbf per inch of width. (N.C.T.E.C., 1965)

Solution

The preceding section describes how a belt transmits motion to a pulley.

The effective pull is the same as the effective force, which is the difference between the tensions on the tight and slack sides of the follower pulley. That is,

$$\text{Effective pull} = P_1 - P_2$$

The smaller pulley runs the faster, and

$$\text{Belt speed} = 2\pi r n_m = 2 \times \frac{22}{7} \times \frac{12}{12} \times 250$$

or,

$$500\pi \text{ ft/min}$$

$$\text{Effective belt tension} = P_1 - P_2 \quad \text{and} \quad P_1 = 2{\cdot}5P_2$$

$$\therefore \text{Effective belt tension} = 2{\cdot}5P_2 - P_2 = 1{\cdot}5P_2$$

$$\text{Horsepower} = \frac{1{\cdot}5P_2 \times 500\pi}{33{,}000} = 30$$

whence

$$P_2 = \frac{30 \times 33{,}000 \times 7}{500 \times 1{\cdot}5 \times 22} = 420 \text{ lbf}$$

and

$$P_1 = 420 \times 2{\cdot}5 = 1{,}050 \text{ lbf}$$

$$\text{Belt width} = \frac{\text{Maximum tension}}{\text{Tension per inch}} = \frac{1{,}050}{120} = 8{\cdot}75 \text{ in.}$$

Example

A belt transmits 4 kW to a pulley 40 cm in diameter which is keyed to a shaft running at 180 rev/min. If the tension of the tight side of the belt is $2\frac{1}{2}$ times the tension of the slack side, calculate the torque on the shaft.

What will be the width of the belt if each 1 cm of its width will safely transmit a pull of 10 kgf.

Solution

$$\text{Let } P_1 = \text{Tension in belt on tight side.}$$

$$P_2 = \text{Tension in belt on slack side.}$$

$$\text{Effective torque} = (P_1 - P_2)r$$

$$\text{Work done by torque per second} = (P_1 - P_2)r\omega$$

$$= \text{Power}$$

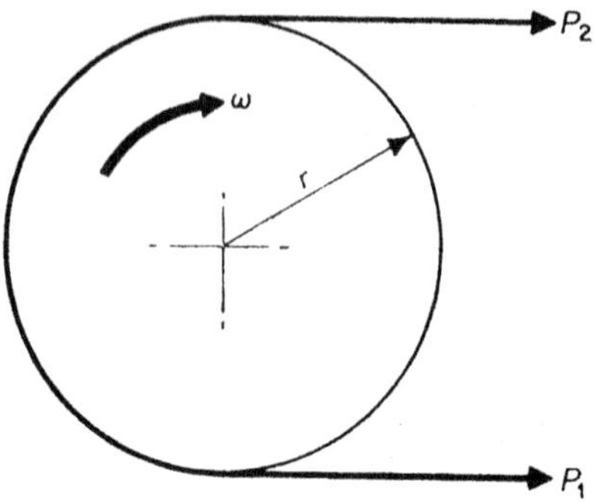

Figure 7.29 Belt driving pulley.

$$\therefore\ 4{,}000 = \left(P_1 - \frac{2}{5}P_1\right) \times \frac{20}{100} \times \frac{180}{60} \times 2\pi$$

$$\frac{3}{5}P_1 = \frac{20{,}000}{6\pi}$$

$$P_1 = 1{,}768 \text{ N}$$

and

$$P_2 = \frac{2}{5} \times P_1 = \frac{2}{5} \times 1{,}768$$

$$= 707 \text{ N}$$

$$\text{Effective torque} = (1{,}768 - 707) \times \frac{20}{100} \text{ Nm}$$

$$= 212 \text{ Nm}$$

$$\text{Width of belt} = \frac{1{,}768 \text{ N}}{9{\cdot}81 \dfrac{\text{N}}{\text{kgf}}} \times \frac{1 \text{ cm}}{10 \text{ kgf}}$$

$$= \frac{1{,}768}{98{\cdot}1} = 18 \text{ cm}$$

Plate (or disc) clutch

A clutch is used when motion is to be transmitted from one shaft to another for intermittent periods. The functions of a clutch are (1) to increase gradually the angular velocity of the driven shaft until it reaches that of the driving shaft, and (2) to rotate the two shafts with the same angular velocity, that is, without slip.

The simplest form of plate clutch consists of two flat plates pressed against each other by means of a spring. If one of the two surfaces which are pressed together has a high coefficient of friction, slip can be prevented. The surface with a high coefficient of friction is usually in the form of an annulus as in figure 7.30.

Let r_m ft be the mean radius of the surface and μ be its coefficient of friction. If P lbf is the force with which the two surfaces are being pressed together, then μP is the total tangential force opposing the rotation.

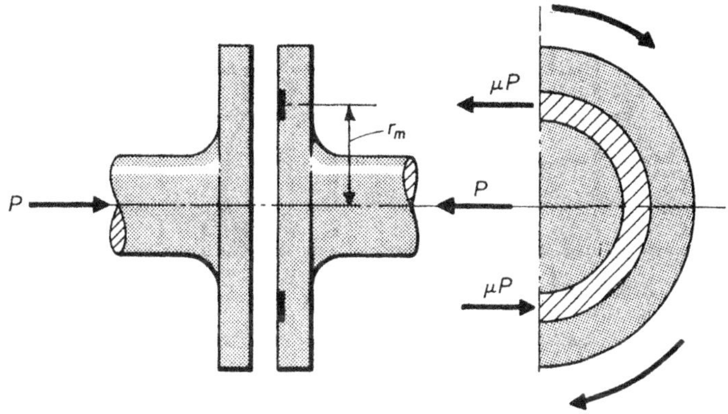

Figure 7.30 A simple plate clutch.

$$\text{Total frictional torque} = \mu P \times r_m \text{ lbf ft}$$

Provided the frictional torque is greater than the resisting torque on the output shaft, the two halves of the clutch, when engaged, will rotate together after a very short interval of time.

Friction brakes

When determining the output of an engine in a laboratory it is convenient to use a form of friction brake to absorb the power in the output shaft. Two types of friction brake are in common use. They are the Rope Brake and the Prony Brake, and in each type friction is applied to the rim of a flywheel or pulley keyed to the engine output shaft.

(*a*) *Rope Brake*
The Rope Brake is the simplest form of brake used, and yet it is suitable for absorbing large power. It is shown diagrammatically in figure 7.31.

In this type of brake a rope is wound once round the circumference of the flywheel, one end of the rope being loaded vertically downwards by a carrier with weights, the other end being attached to a spring balance which is secured to an overhead point immediately above the hanging weights. Note that rotation of the flywheel tends to lift the carrier and the weights.

If a positive reading is obtained from the spring balance, the

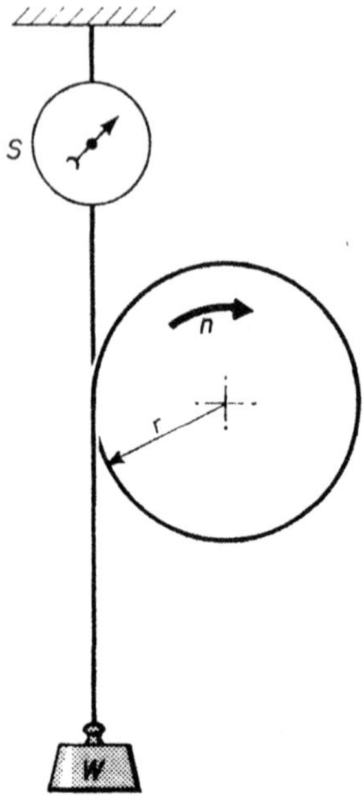

Figure 7.31 The rope brake.

clockwise engine torque is balanced by the net anti-clockwise torque of the rope, thus

$$\text{Net, or effective torque} = (W - S)r \text{ lbf ft}$$

where r = Mean distance of the centre of the rope from the axis of rotation.

$$\text{Work done by this torque} = (W - S)r \times 2\pi n_m \text{ ft lbf/min}$$

where n_m = Revolutions of flywheel per minute.

$$\text{Horsepower absorbed} = \frac{(W - S) \times 2\pi r n_m}{33{,}000}$$

$$\text{Power in watts} = (W - S) \times 2\pi r n_s \text{ (SI units)}$$

Note that the horsepower absorbed, usually termed the "Brake Horsepower" is not dependent on the type of engine nor on the number of cylinders.

(b) Prony Brake

This type of brake is also very simple and is used for absorbing small powers, the flywheel or pulley being only about 4 to 6 inches in diameter.

The brake consists of two wooden blocks with a central hole fitting over the pulley in such a manner that the frictional torque can be adjusted by tightening or slackening wing nuts (see figure 7.32).

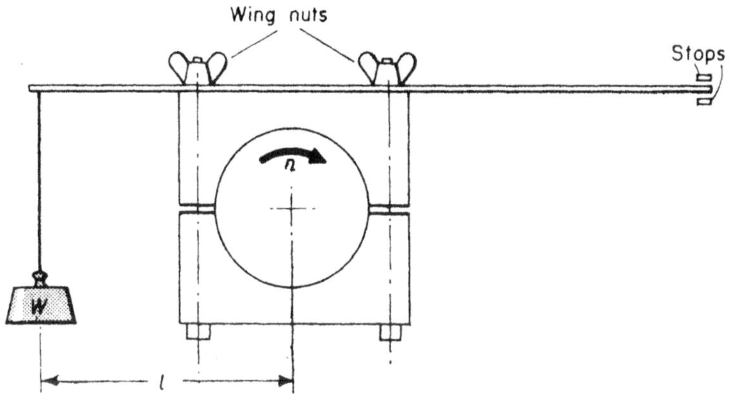

Figure 7.32 The prony brake.

A lever secured to the top surface of the upper block carries a load of W lbf at its free end. At the other end it is restricted in its vertical movement by two stops.

When the engine is running at a constant speed the wing nuts should be adjusted to give a constant braking torque when a load of W lbf is hanging at the free end of the lever. The wing nuts are adjusted so that the lever floats between the two stops, the anti-clockwise moment of the load W being balanced by the clockwise turning moment at the circumference of the pulley.

If l ft = horizontal distance of the weight W lbf from axis of pulley,

$$\text{Resisting moment} = \text{Torque} = W \times l \text{ lbf ft}$$

$$\text{Work done by this torque} = Wl \times 2\pi n_m \text{ ft lbf/min}$$

where n_m = Revolutions of flywheel per minute.

$$\therefore \text{ Horsepower absorbed} = \frac{Wl \times 2\pi n_m}{33{,}000}$$

$$\text{Power in watts} = Wl \times 2\pi n_s \text{ (SI units)}$$

(where n_s represents revolutions per second.)

Example

A car of mass 1 t climbs a gradient of 1 in 100 at a constant speed of 48 km/h. If the engine is working at 14·5 kW, calculate the frictional resistance.

What would be the maximum velocity of the car on the level if the frictional resistance and horsepower remained the same? (U.L.C.I.)

Solution

(*a*) Work output of engine = Work to lift car + Frictional resistance

whence

Work against frictional resistance = Work of engine − Work to lift car

$$\text{Engine work per second} = 14{\cdot}5 \times 1000\ \text{J} = 14{,}500\ \text{J/s}$$

$$\text{Work to lift car} = 1000\ \text{kg} \times 9{\cdot}81\ \text{N} \times \text{Distance vertically per second}$$

$$= 9810 \times \frac{4800}{3600} \times \frac{1}{100}$$

$$= 1308\ \text{J/s}$$

$$\therefore\ \text{Friction work} = 14{,}500 - 1308 = 13{,}192\ \text{J/s}$$

Friction acts along the slope, and is overcome at the car speed.

$$\therefore\ \text{Friction resistance} \times \frac{48{,}000}{3600} = 13{,}192\ \text{J/s}$$

and

$$\text{Frictional resistance} = \frac{13{,}192 \times 36}{480} = 989{\cdot}4\ \text{N}$$

(*b*) Let

$$\text{Maximum velocity on level} = v\ \text{ft/s}$$

Then

Frictional resistance × Speed on level = Engine output

$$\therefore\ 989{\cdot}4\ \text{N} \times v = 14{,}500$$

$$v = \frac{14{,}500}{989{\cdot}4} = 14{\cdot}65\ \text{m/s}$$

$$\text{Maximum speed on level} = \frac{14{\cdot}65 \times 3600}{1000} = 52{\cdot}74\ \text{km/h}$$

Summary

Mechanical advantage is defined as the ratio:

$$\frac{\text{Load}}{\text{Effort}}$$

Velocity ratio is given by the relationship:

$$\frac{\text{Distance moved by effort}}{\text{Distance moved by load}}$$

Efficiency is given by:

$$\frac{\text{Work output}}{\text{Work input}}$$

which can be expanded as:

$$\text{Efficiency} = \frac{\text{Mechanical advantage}}{\text{Velocity ratio}}$$

The law of a machine
This is of the form:

$$P = mW + c$$

where P is the effort and W is the load. This is a linear relationship, and m is the slope of the line.

The wedge
For a wedge of angle 2θ, driving force P_d, and slit normal reaction R, the following relationships apply:

With no friction,

$$\text{Mechanical advantage} = \frac{1}{2 \tan \theta}$$

With friction,

$$\mu R = \frac{P_d}{2(\sin \theta + \mu \cos \theta)}$$

For withdrawing,

$$\text{Force } P_w = P_d \left(\frac{\mu \cos \theta - \sin \theta}{\sin \theta + \mu \cos \theta}\right)$$

Pulleys with continuous cords
In the simple pulley block without friction, the cord tension is everywhere the same, and the effort is equal to the load.

In other pulley blocks where the cord is continuous between the point of application of the effort and the final anchorage, and friction is neglected, the cord tension is again everywhere the same, and is equal to the effort.

The mechanical advantage is evidently the same as the number of cords supporting the load.

Pulleys with discontinuous cords

Whenever a pulley is suspended in a loop which is shortened, the pulley rises half the amount of the loop shortening.

The Velocity Ratio of the system is unaffected by friction.

Weston's pulley block

$$\text{Velocity ratio} = \frac{2 \times \text{large dia. of differential wheel}}{\text{Diff. of diameters of differential wheels}}$$

Overhauling

If, when the effort is removed, the load overcomes the frictional resistance, and begins to fall, the machine is said to overhaul.

If overhauling is to be avoided the efficiency of the machine must be less than 50 per cent.

Screw threads

These can be considered as elementary machines. If the thread angle is θ, the lead is l, and the mean diameter of the thread is d, then

$$\theta = \tan^{-1} \frac{l}{\pi d}$$

$$\text{Mechanical advantage} = \frac{W}{P}$$

$$= \left(\frac{\cos\theta - \mu \sin\theta}{\mu \cos\theta + \sin\theta}\right)$$

$$\text{Velocity ratio} = \frac{1}{\tan\theta}$$

The screw jack

In this mechanism the effort is applied at the end of a lever, and the load rises with the screw.

Power

This is defined as the rate of doing work.

The mechanical unit of power in the FPS system is the horsepower. The power unit on the SI system is the joule per second, or watt.

1 horsepower is equivalent to a working rate of 33,000 ft lbf per minute.

Work done by a torque

This is given by the product of the torque and the angle turned through, in radians.

For a torque T rotating through n_m revolutions each minute

$$\text{Horsepower} = \frac{2\pi n_m T}{33{,}000} \quad \text{and} \quad \text{Power (watts)} = 2\pi n_s T \text{ (SI units)}$$

Clutches

Here the torque is given as

$$T = \mu P \times r_m$$

where μ = Coefficient of friction,
P = Force engaging clutch, and
r_m = Effective radius.

Rope brake

In the rope brake the effective force at the periphery is given by $(W - S)$, where W is the load suspended on the rope, and S is the spring tension.

$$\text{Horsepower} = \frac{(W - S) \times 2\pi r n_m}{33{,}000}$$

Prony brake

In the Prony brake the torque is measured by a load W acting at a distance l from the centre of rotation, to give a torque Wl.

$$\text{Horsepower} = \frac{Wl \times 2\pi n_m}{33{,}000}$$

and $$\text{Power in watts} = Wl \times 2\pi n_s \text{ (SI units)}$$

Exercises

1. The law of a lifting machine is given by

$$P = 0{\cdot}25W + 0{\cdot}15 \text{ kN}$$

If the velocity ratio of the machine is 5, calculate the effort required to raise a load of 6 kN and the actual efficiency at this load.

2. The following table gives values of load and effort obtained during an experiment on a certain machine.

Load (Wkg)	10	20	30	40	50	60
Effort (PN)	123	162	206	240	275	319

Plot a graph and determine the law of the machine.

3. The angle of a wedge is 10°. A force of 20 lbf is applied along its axis. Find the forces at right angles to the surface of the wedge.

4. A force of 2·94 kN is applied axially to a wedge which has an angle of 7°. Find the splitting force, normal to the wedge axis, produced by this force.

5. A wedge is used for forcing apart two flat surfaces. The taper of the wedge is 1 inch in each foot of its axial length. If a force of 1000 lbf is required to force the surfaces apart, and the coefficient of friction between the wedge and the two surfaces is 0·25, find

(a) the forces normal to the wedge surfaces resisting separation,

(b) the force required to withdraw the wedge before the surfaces have separated.

6. A load is raised by means of a pair of pulley blocks, the upper one having three pulleys and the lower one having two. Find

(a) the velocity ratio

(b) the effort required to lift a load of 600 kg if the mass of the lower block is 10 kg, and the efficiency is 80 per cent at this load

(c) the mechanical advantage.

7. A lifting machine has a velocity ratio of 5 and can lift a load of 200 lbf when an effort of 50 lbf is applied. Determine

(a) the efficiency

(b) the effort required to overcome friction

(c) the work done against friction when the load is raised 1 ft.

8. A load of 500 lbf is attached to a cable which is wrapped round the axle of a simple wheel and axle, the diameters of which are 2 ft and 8 in. respectively. The effort is applied to a cable which is wrapped round the wheel. If the efficiency of this machine is 70 per cent, calculate

(a) the effort required to raise the load

(b) the effort required to prevent the load from falling.

9. When the load on a certain lifting machine is increased to 500 kg the machine begins to overhaul. If its velocity ratio is 10, calculate the effort required to raise this load.

10. A 6 in. diameter shaft is rotating in a journal bearing at 200 rev/min. The load on the journal is 4 tonf and the coefficient of friction between the shaft and its bearing is 0·05. Calculate

(a) the frictional torque

(b) the horsepower absorbed in friction.

11. Calculate the frictional torque resisting the rotation of a 4 in. diameter shaft in a journal bearing if the load on the bearing is 2 tonf and the coefficient of friction between the shaft and the bearing is 0·03.

If the shaft is rotating at 250 rev/min what horsepower is being absorbed in friction?

12. The speed of a belt is 2000 ft/min and it transmits 50 hp. If the tension on the tight side of the belt is $2\frac{1}{2}$ times that on the slack side, calculate the tensions on the two sides.

13. The tension in the tight side of an endless belt is 3 times the tension in the slack side. The driving pulley has a diameter of 30 in. If 3 hp is being transmitted when the pulley is rotating at 250 rev/min determine the tensions in the belt.

14. Calculate the efficiency of a screw jack in which the screw has a single start thread of $\frac{1}{4}$ in. pitch and a diameter of 2 in. The effort is applied at the end of a lever 10 in. from the axis of the jack to lift a load of 1000 lbf, and the coefficient of friction is 0·15.

15. A screw jack has a single start thread of $\frac{1}{4}$ in. pitch. Effort is applied at one end of a tommy bar to raise a load of 2 tonf. If a force of 30 lbf applied at a distance of 15 in. from the axis of the screw is sufficient to raise the load, calculate the efficiency of the jack at this load.

16. A force of 392 N is required to lift a load of 3500 kg by means of a screw jack with a single start thread. If the effort is applied at a distance of

50 mm from the axis of the jack, and its efficiency is 40 per cent, what is the pitch of the thread to the nearest millimetre?

17. Derive an expression for the velocity ratio of a Weston differential pulley block.

A differential block has 10 flats on the large pulley and 9 on the small pulley. Determine the effort required to raise a load of 360 lbf if the efficiency at this load is 35 per cent.

Also determine the work done against friction when this load is raised 1 ft. (E.M.E.U.)

8

Equilibrium of forces and plane frameworks

Equilibrium means a state of balance.

When a rigid body is at rest, under the action of a number of external forces all of which are acting in one plane,

1. there is no tendency for the body to move in a straight line in any direction, and
2. there is no tendency for the body to rotate.

If the body is not to move laterally all the forces acting on it must balance; that is, in the vertical direction, the upward forces must balance the downward forces and, in the horizontal direction, forces to the left must balance forces to the right. If the forces are inclined to these two directions, they must first of all be resolved vertically and horizontally, or along any other two directions at right angles to each other.

If the body is not to rotate, then clockwise moments must balance anti-clockwise moments about any point; that is the algebraic sum of the moments of all the forces about any point in the plane must be zero.

These are the general conditions for equilibrium and must be thoroughly understood and memorized.

Two opposing forces

If there are only two forces in equilibrium these two forces must be equal and opposite, and further, must act in the same straight line, as in figure 8.1.

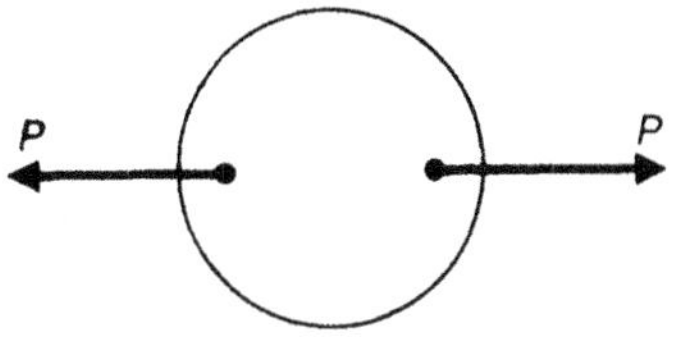

Figure 8.1 Two equal and opposite forces.

If the forces are equal in magnitude, but are not originally acting in the same straight line, the body will rotate until the two lines of action are in the same straight line—figure 8.2.

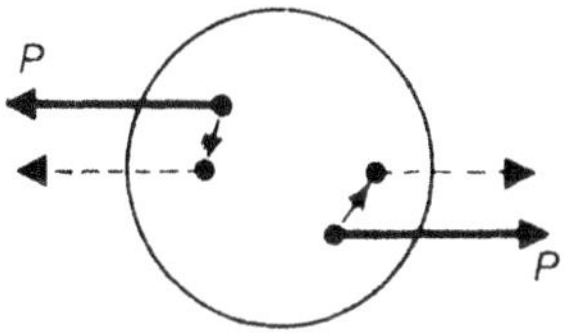

Figure 8.2 Two equal and opposite forces, not in the same straight line.

Three parallel forces

Suppose three forces are in equilibrium, and their lines of action are

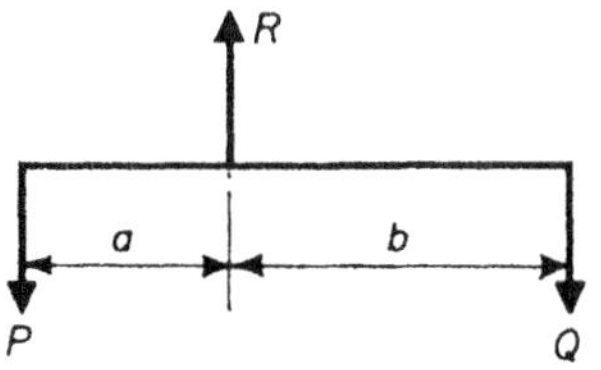

Figure 8.3 Three parallel forces.

parallel. Referring to figure 8.3, in order to satisfy the two conditions of equilibrium.

$$\text{Upward force} = \text{Sum of downward forces}$$

or,

$$R = P + Q$$

and

$$\text{Clockwise moments} = \text{Anti-clockwise moments}$$

or, taking moments about the left hand end

$$Q(a + b) = Ra$$

Three non-parallel forces

If the three forces are not parallel, their lines of action must meet in a single point if equilibrium is required—figure 8.4.

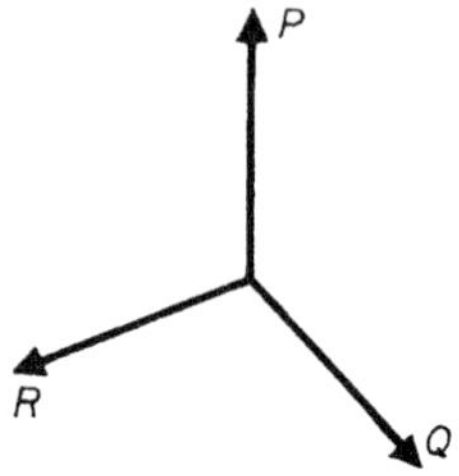

Figure 8.4 Space diagram.

The force diagram takes the form of a triangle, and, for equilibrium the triangle must close, as in figure 8.5. That is, there is no resultant force.

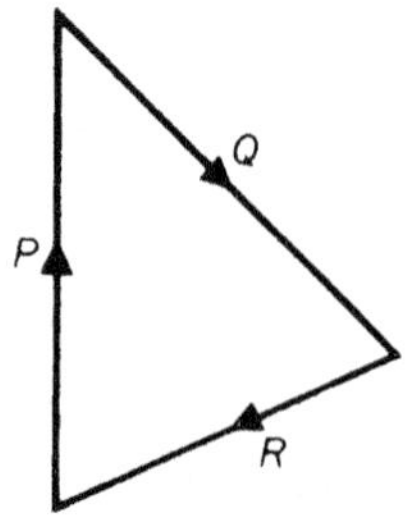

Figure 8.5 Force diagram.

Reactions

Surfaces are stated in mechanics to be either smooth or rough. In practice all surfaces have some degree of roughness.

Smooth surface

If the surface is smooth, and a load rests on it, the reaction of the surface to the load is always normal to the surface.

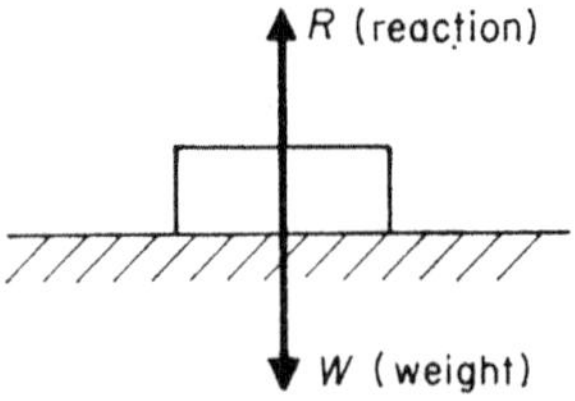

(a)

Figure 8.6(*a*) Smooth horizontal surface.

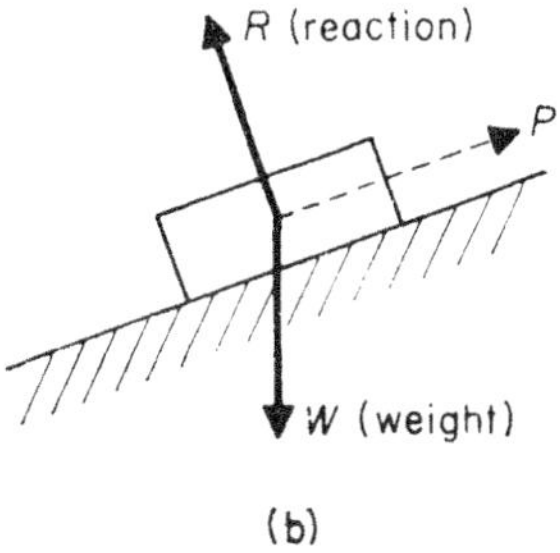

Figure 8.6(*b*) Smooth inclined surface.

Rough surface
If the surface is rough, and an inclined force is exerted on it, the direction of the surface reaction is *NOT* normal. Fundamentally, there are two forces acting at the surface—(figure 8.7)—when an inclined force P is applied. These two forces consist of one normal to the surface and a frictional force tangential to the surface resisting slipping. When combined they produce a resultant reaction which is not normal to the surface but is equal and opposite to the applied force.

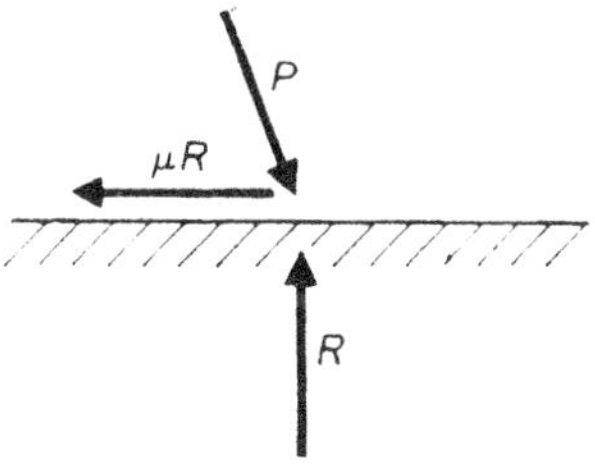

Figure 8.7 Rough surface.

Consider a ladder resting on a rough horizontal floor, and leaning against a smooth vertical wall as in figure 8.8.

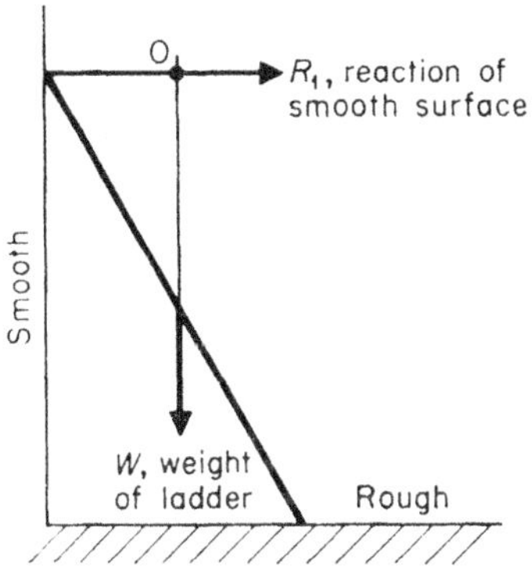

Figure 8.8 Ladder leaning against a wall.

Three forces are involved. One force—the weight of the ladder—is known in magnitude and direction. Because the vertical wall is smooth, its reaction to the ladder must be normal to the wall and therefore the direction of R_1 is known. The resultant reaction of the rough floor is not known, either in direction or magnitude.

From the condition for equilibrium of three forces their lines of action must meet in a single point. This point, must be the intersection of the lines of action of the weight and the normal reaction to the wall. Consequently, if the space diagram is drawn to scale, it being assumed that the weight W acts through the mid-point of the length of the ladder, the point of intersection, O, of the weight W and the reaction R_1 can be obtained.

Because the line of action of the third force, the reaction R_2 of the floor, must pass through O, join O to the point of contact of the ladder with the floor, thus giving the line of action of R_2—(figure 8.9).

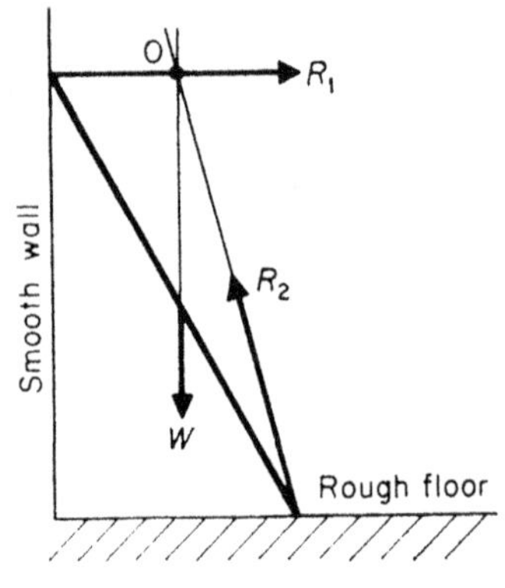

Figure 8.9 The line of action of the floor reaction.

Having determined the line of action of R_2 from the space diagram, drawn to scale as in figure 8.9, its magnitude can be found by drawing a force diagram. Because the magnitude of W is known, this value determines the scale of the diagram, and the sides of the triangle in figure 8.10 are drawn parallel to the three forces. The sense of W determines the sense of the other two forces since it is necessary to proceed round the triangle in the direction indicated by the arrow for force W.

Example

A ladder of uniform construction is 20 ft long and weighs 60 lbf. One end of the ladder is placed against a smooth vertical wall; the other end rests upon a rough horizontal surface at a point which is 5 ft from the

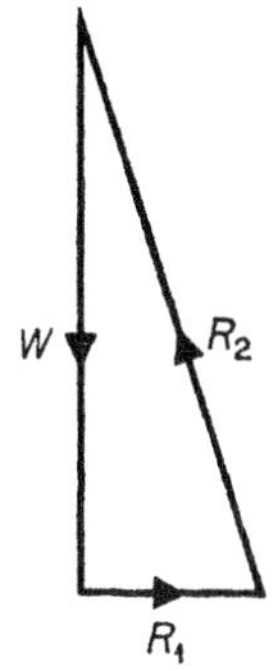

Figure 8.10 Force diagram to scale.

foot of the wall. Determine the magnitudes and directions of the reactions of the wall and the ground.

Solution

Draw to scale the ladder leaning against a vertical wall, as in figure 8.11(*a*). The reaction R_1 must be perpendicular to the wall because the wall is smooth.

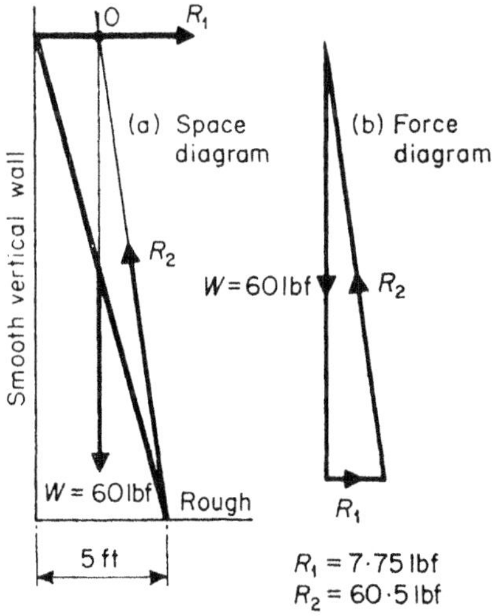

Figure 8.11

Produce the line of action of the load W, a vertical line through the mid-point of the ladder, until it meets the line of action of R_1 at O. Then, because the external forces acting on the ladder are in equilibrium and therefore meet in a single point, the resultant reaction of the floor

R_2 must pass through O. This determines the direction of R_2. Knowing the line of action of R_2 from the space diagram its magnitude is found by drawing a force diagram to scale. Because the magnitude of W is known, this value determines the scale of the diagram. The sides of the diagram in figure 8.11(*b*) are then drawn parallel to the three forces acting on the ladder, giving the magnitudes of R_1 and R_2 to the same scale as W.

The direction of W determines the direction of the other two forces by proceeding round the force diagram in the direction indicated by the arrow for force W—that is, in an anti-clockwise direction.

Hinge reaction

If a force is applied through a hinged joint the reaction is *NOT* normal to the fixed surface.

A trap door of weight W is hinged at one end to a vertical wall, and

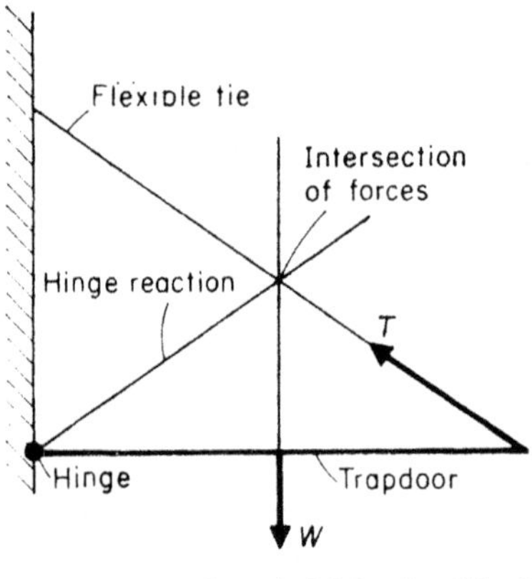

Figure 8.12 Trapdoor held by flexible tie.

held in position by a flexible tie secured at one end to the trapdoor and at the other to a point in the wall vertically above the hinge.

Figure 8.12 shows such a trapdoor held in a horizontal position—although this is not essential.

If the door is of uniform construction, assume that its weight W will act vertically downwards through a point mid-way between the hinge and the free end.

Because the tie is flexible, the tension (T) in it must be axial, and therefore the point of intersection of the lines of action of W and T is readily determined. From the general conditions of equilibrium, the third force—which is the reaction of the hinge—must pass through this point of intersection. Therefore, if the space diagram is drawn to scale the inclinations of the lines of action of the three forces are determined as in figure 8.12.

Example

A trapdoor is hinged at one end to a vertical wall and supported at the other with a rope 2·04 metre long secured to a ring in the wall at a point 1·60 metre vertically above the hinge. The trapdoor is 1·28 metre long and weighs 50 N. Determine the tension in the rope and the reaction of the hinge.

Solution

Draw the space diagram to scale and determine the point O, where the lines of action of the weight (W) of the trapdoor and the tension (T) in the rope intersect. Join O to the hinge, to give the line of action of the hinge reaction.

Select a suitable scale for the force diagram and draw a vertical line

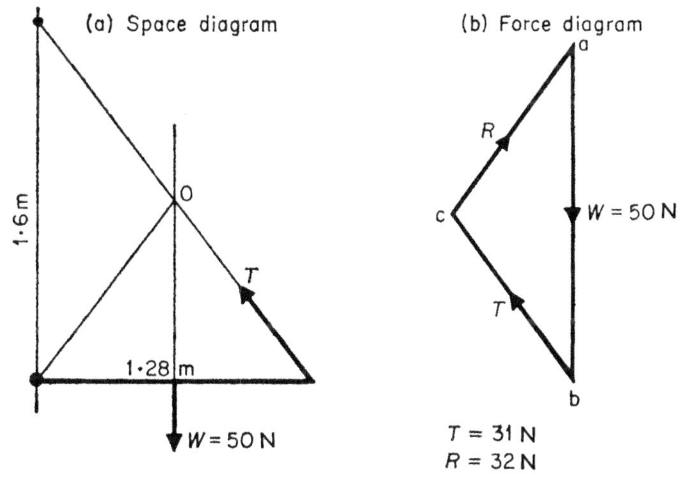

Figure 8.13(*a*) Space diagram; (*b*) Force diagram.

ab to represent the weight W. Draw lines parallel to the lines of action of the rope tension and the hinge reaction to complete the force diagram as shown in figure 8.13(*b*). Then the lines *bc* and *ca* represent T and R to the same scale as *ab* represents W.

Suggested scales are:

(*a*) 1 cm represents 0·25 metre.
(*b*) 1 cm represents 5 N.

From the force diagram

$$T = 31 \text{ N}$$

$$R = 32 \text{ N}$$

Co-planar concurrent forces

When the lines of action of a number of forces lie in one plane and pass through a common point they are termed co-planar concurrent forces.

It will be understood that if a system of forces is in equilibrium—as at a joint in a structural framework—then the force diagram must be a closed figure. This is the basis for the graphical analysis of frames.

In practice in a framework two forces would normally provide the equilibrant. The example below illustrates this.

Example

A structural joint has six members which radiate from a common point.

The line of action of one member is vertically upwards and exerts a force of 3 tonf pulling away from the common point. The other members

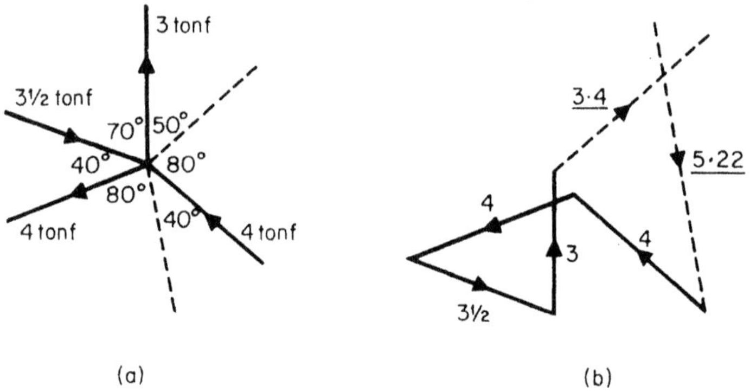

Figure 8.14

taken in clockwise rotation from the vertical member are at angles 50°, 130°, 170°, 250°, and 290° to it. The forces exerted at the common point are respectively, unknown; 4 ton pushing; unknown; 4 ton pulling; 3½ ton pushing.

Find the magnitude and sense of the unknown forces in the two members. (U.E.I.)

Solution

Starting with the 4 ton pushing force, draw the force vector to a suitable scale, at 130° to the vertical. Continue in a clockwise direction, adding in turn the vectors representing the other three known forces.

From the extremities of this vector diagram construct lines parallel to the directions of the two unknown forces. Continue these lines until they intersect, so that a closed figure is produced. The two closing lines are vectors representing completely the two unknown forces.

The unknown forces in the 50° and 170° members are determined as 3·4 tonf pulling and 5.22 tonf pulling respectively, as in figure 8.14(*b*).

When the forces of a system are *concurrent*, the resultant force must pass through the point which is common to all the lines of action. The resultant can be determined completely from the construction of the force diagram.

Co-planar non-concurrent forces

If the forces do not pass through a common point then the system is called *non-concurrent*. The force diagram may still be used to determine the magnitude, direction and sense of the resultant but one additional

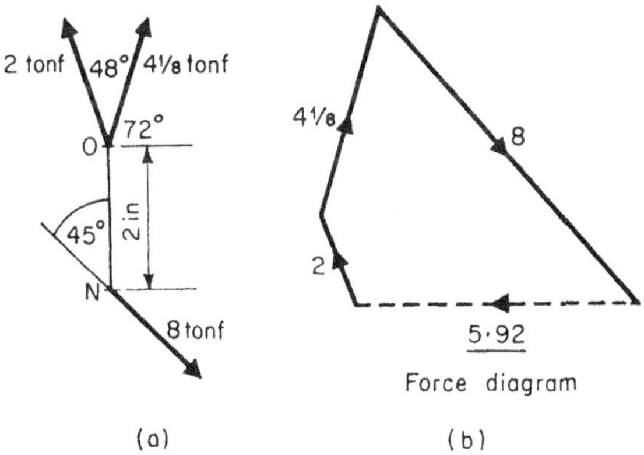

Figure 8.15

step is required to find its line of action. A graphical process is available for this step, but, at this stage, it is considered advisable to rely on an analytical method. The basis of this method is that, *about any reference point, the moment, or rotational influence, of the resultant must be equal to the sum of the moments of the individual forces.*

The conditions for the equilibrium of a system of co-planar, non-concurrent forces, are

1. the force diagram must close, and
2. the resultant turning moment about any point must be zero.

Example

Draw a force diagram for the system of forces shown in figure 8.15 and from it obtain the magnitude, direction and sense of the equilibrant.

Obtain the position in space of the equilibrant by taking moments of the force about O. (U.E.I.)

Solution
From the force diagram, the equilibrant is determined as a force of 5·92 tonf acting at right-angles to ON and in the sense indicated, that is, to the left. Taking moments about O,

$$5{\cdot}92 \times y = 8 \times 1{\cdot}414$$
$$y = 1{\cdot}91 \text{ in.}$$

The equilibrant acts through a point 1·91 in. from O measured along ON.

Co-planar parallel forces

In engineering practice it is quite usual for a structure, or a member of a structure, to be subjected to a system of forces having parallel lines of action. Examples of this are the girders of a bridge, the floor beams in a building, or a complete roof truss.

Although a number of parallel forces represents a special case of a non-concurrent force system, the analysis of such a system and the conditions for equilibrium are unchanged. The special circumstances do, however, allow simplification of the force diagram; instead of being an open figure, having two dimensions, the diagram is a single straight line.

Parallel forces are sometimes referred to as like or unlike forces according to whether they are of the same or different sense. This is not an important distinction if the principle of sense, as applied to the taking of moments, is understood. A symbolic notation, as illustrated in the following example, may be helpful in this respect.

Example

Four parallel loads, as shown in figure 8.16, act on a member. Determine the resultant of these and hence the equilibrant. Forces are given in kilonewtons and distances in metres. Note: 1 kilonewton = 1000 newtons.

Solution
Starting with the 6 kN load and working in a clockwise direction construct the appropriate force diagram as in figure 8.16(*b*). From this diagram it is clear that the resultant is a 10 kN force acting downwards, and that it is in the same sense as the 6, 2 and 5 kN forces. The equilibrant, by definition, must be equal and opposite to the resultant.

To determine the line of action of the equilibrant. Moments may be taken about any point. Use a point P, 2 m to the left of the 6 kN force, as shown.

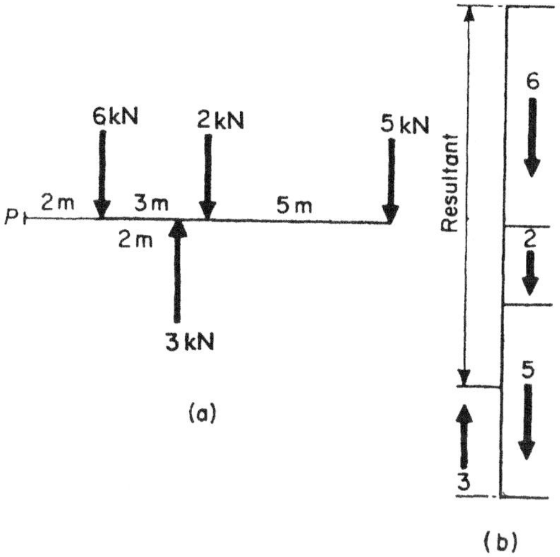

Figure 8.16

Taking moments about P,

Clockwise moment↻ = (6 kN × 2 m) + (2 kN × 5 m) + (5 kN × 10 m)
= 72 kilonewton-metres

Anticlockwise moment↺ = 3 kN × 4 m
= 12 kN-m

Moment of resultant = 72 kN-m↻ − 12 kN-m↺
= 60 kN-m↻

Therefore

Moment of equilibrant = 60 kN-m↺

Since it has been established that the equilibrant force is 10 kN upwards then its line of action must be 6 m from P, and to the right of it in order to provide an anticlockwise moment of 60 kN-metres. Its position is shown in figure 8.17.

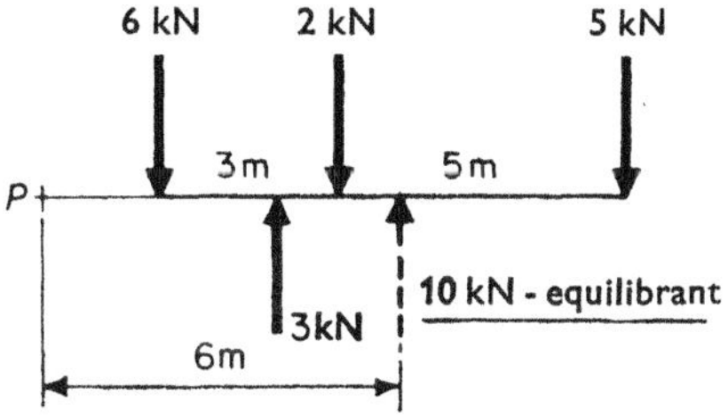

Figure 8.17

The arithmetic would have been simplified slightly if moments had been taken about a point on the line of action of one of the forces, say the 6 kN, instead of about an arbitrary point such as P; the calculation is as follows:

Clockwise moment ↻ $= 2\text{ kN} \times 3\text{ m} + 5\text{ kN} \times 8\text{ m}$
$= 46\text{ kN-m}$

Anticlockwise moment ↺ $= 3\text{ kN} \times 2\text{ m}$
$= 6\text{ kN-m}$

Moment of resultant ↻ $= 46\text{ kN-m}$ ↻ $- 6\text{ kN-m}$ ↺
$= 40\text{ kN-m}$ ↻

Moment of equilibrant $= 40\text{ kN-m}$ ↺

Being a 10 kN force, the equilibrant must act at a distance of 4 m from the point about which moments have been taken.

This result is consistent with the one derived earlier.

In this example the magnitude and line of action of a single force—the equilibrant—have been determined.

A more common requirement in a practical example is the determination of two forces which together provide the equilibrant; often their lines of action are known. To do this it is necessary to carry out the procedure of taking moments twice, once each about a point in the line of action of the two unknown forces. A rearrangement of the previous problem will show the method.

Example

A beam 8 metres long carries vertical loads of 6, 2 and 5 tonnes at distances 0, 3 and 8 metres respectively from the left hand end. It is supported at two points, 2 metres and 4 metres from the left hand end.

Calculate the magnitude of the reaction at each support.

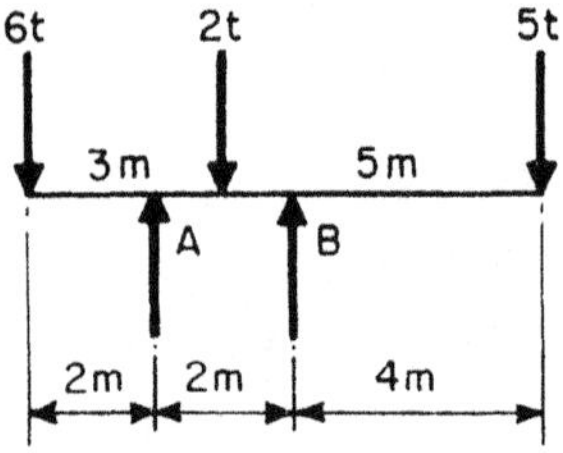

Figure 8.18

Solution

Let the points of support be A and B and the reaction forces R_A and R_B respectively. Ignore the 9810 N/t factor until later.

Taking moments about A,

For equilibrium, anticlockwise moments = Clockwise moments

$$\circlearrowleft (R_B \times 2) + (6 \times 2) = (2 \times 1) + (5 \times 6) \circlearrowright$$
$$(R_B \times 2) + 12 = 2 + 30$$
$$R_B \times 2 = 20$$
$$R_B = 10 \text{ t} \uparrow \ (\times\ 9{\cdot}81 \text{ kN/t})$$

Taking moments about B,

$$\circlearrowleft (R_A \times 2) + (5 \times 4) = (6 \times 4) + (2 \times 1) \circlearrowright$$
$$(R_A \times 2) + 20 = 24 + 2$$
$$R_A \times 2 = 6$$
$$R_A = 3 \text{ t} \uparrow \ (\times\ 9{\cdot}81 \text{ kN/t})$$

Check. All external forces must balance,

$$R_A + R_B = 6 + 2 + 5 = \text{sum of loads}$$

Bow's notation

When more than three forces are involved, and it is necessary to draw both a space diagram and a force diagram, it is useful to 'letter' the diagrams in a manner known as 'Bow's Notation'.

In this notation the *spaces* between adjacent forces on a space

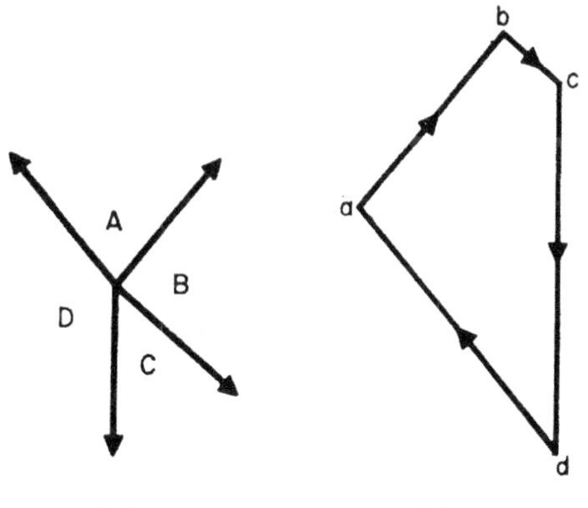

(a) (b)

(*a*) Space diagram (*b*) Force diagram

Figure 8.19 Bow's notation.

diagram are identified by capital letters as in figure 8.19 instead of giving letters, or numbers, to the forces themselves. This should be done in a regular manner, in either a clockwise or an anticlockwise direction.

In the force diagram small letters are used, and the ends of each force in this diagram are given the same letters as those in the spaces on either side of the corresponding force in the space diagram.

The force diagram, of course, must be drawn to scale, and the scale used must be stated. Further, to proceed round the force diagram in a clockwise direction as with Bow's Notation in the space diagram, *a* to *b*, and NOT *b* to *a* must represent the direction (or sense) of the force AB.

In the simple case shown in figure 8.19 four 'pull' forces are acting at a single point. First draw force *ab* parallel to the direction of AB in the space diagram, and to a convenient scale. It is then only necessary to continue round the space diagram in a clockwise manner, and to draw *bc* to represent BC, *cd* to represent CD, and *da* to represent DA, the directions in the diagram being parallel to the respective lines of action of the forces in the space diagram.

If the four forces are in equilibrium the force polygon must close. That is, there will be no resultant force.

Simple framed structures

Many stationary and mobile structures, such as roofs, bridges and cranes, are often built up from components joined together to form triangular shapes. A structure made up of four-sided frames would not be stable, because a load applied at one of the corners would tend to make the frame collapse. A three-sided frame, however, is rigid and cannot change its shape unless the components, or structural members, distort.

For the purpose of designing structures built up from triangular frames it is assumed that the joints, or corners of the frames, are not rigidly connected. Such joints are known as pin joints, it being assumed that the members forming the joint can rotate freely relative to one another like a hinge, and without friction. The force in any member is taken to be axial and must therefore be either a tensile or a compressive force.

It is further assumed that the external forces are applied *only* at the joints of the framework.

In most frameworks the direction of the applied external force is not coincident with the axis of any of the structural members which meet at the joint. However, the external force can be replaced by components which do act along the axes of the members, and, because all the forces at a joint are balanced, these components are opposed to, and equal, the internal forces in the members. If the components of the external forces applied to the ends of a member are tensile forces, and for equilibrium they must be equal and opposite, the member is said to be a *tie*.

The forces inside the member itself resisting the external 'pull' forces must also be 'pull' forces, as in figure 8.20(*b*).

Similarly, if the external forces applied to the ends of a member

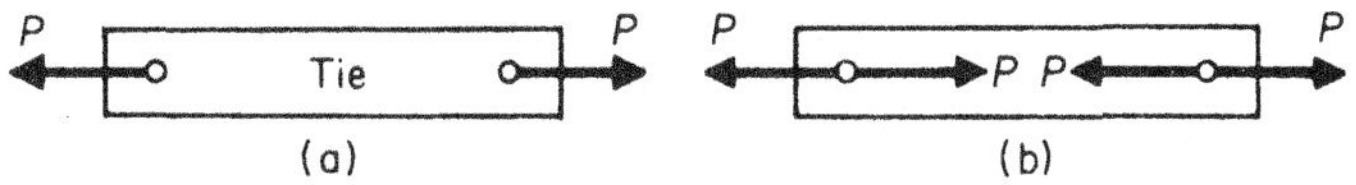

(*a*) Externally applied tensile forces
(*b*) Internal 'pull' forces resisting external 'pull' forces

Figure 8.20 A tie.

are compressive forces, the member is defined as a *strut*, and the forces inside the member resisting compression must be 'push' forces, as in figure 8.21(*b*).

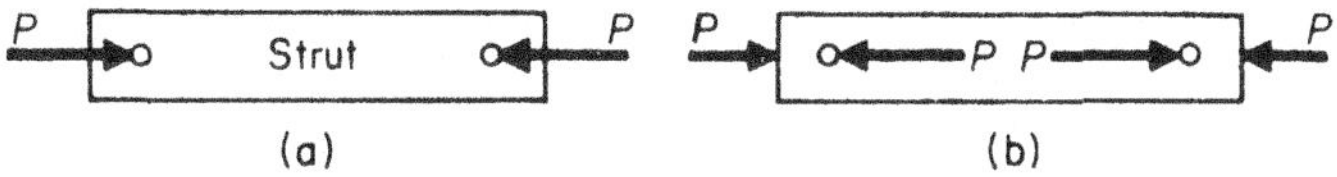

Figure 8.21 A strut.

(*a*) Externally applied compressive forces
(*b*) Internal 'push' forces resisting external 'push' forces

The framed structure will usually rest on two supports and these will normally be situated at the extreme ends of the framework; at least one will be a simple support, that is, one resting on a support without restriction in movement. Sometimes a roller may be inserted between the framework and the support as in figure 8.22(*b*). The second support may be simple or it may be a hinge as in figure 8.22(*c*).

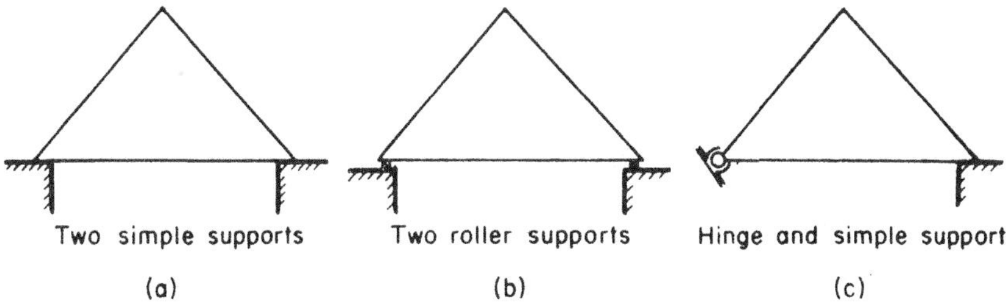

Figure 8.22 Supports for a framed structure.

The reaction at a simple support will be vertical, whereas the reaction at a hinge will be unknown in direction and must be determined.

The first task in any framework problem is to determine the magnitude and direction of the support reactions. Because the framework is in equilibrium—that is, it is stable and stationary under load—the general conditions of equilibrium apply to all the external forces, including the reactions, acting on it.

(a) Forces

All the external forces must balance; that is, the force polygon must close. Consequently, if all the external loads and one reaction are vertical, the second reaction, whether it is a hinge or not, must also be vertical. This is because the force diagram will be a single vertical line.

If all the external loads are not vertical, and one of the supports is a hinge, take moments of all the external forces, including the reactions, about the hinge and hence determine the magnitude of the vertical reaction of the simple support. Then use Bow's Notation and draw the force polygon to determine the magnitude and direction of the hinge reaction.

(b) Moments

The algebraic sum of the moments of all the external forces about any point must be zero. Take moments about each support in turn. When considering the simple framework in figure 8.23 the magnitude of each

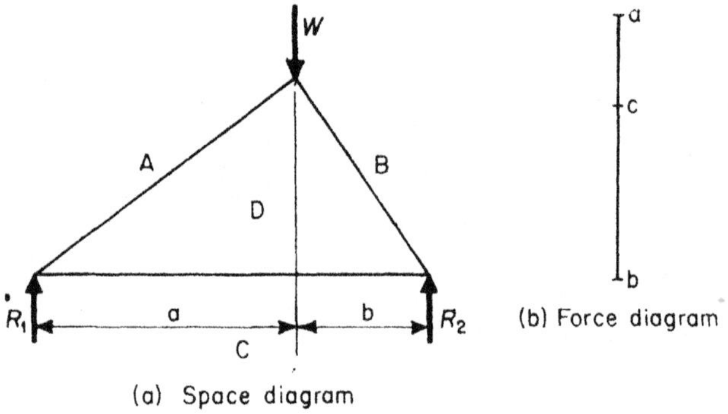

Figure 8.23 Simple framework, simply supported.

reaction can be calculated because the moment of a force about a point on its own line of action is zero, and therefore one of the two reactions will be eliminated from each moment equation in turn.

Taking moments about the left hand support of figure 8.23

$$Wa = R_2(a + b)$$

because the moment of R_1 about its own support is zero.

Similarly, taking moments about the right hand support

$$R_1(a + b) = Wb$$

because the moment of R_2 about its own support is zero.

Having thus obtained both the magnitudes (from the above equations) and the directions (from the force polygon—in this case a straight vertical line) of the two reactions, it is now possible to obtain the axial load in each of the members.

Each joint may be regarded as a point through which a number of forces act, producing a state of equilibrium. These forces may be 'push' forces or 'pull' forces. *For each joint a force polygon can be drawn, provided at least one force is known in magnitude, and not more than two forces are unknown.* Decide on a scale for the force diagram, determined by the magnitude of the known external forces.

Refer again to figure 8.23 and apply Bow's Notation to it; Consider the joint at the left hand support, where the reaction R_1 has been determined, but the magnitude and types of force in members AD and DC are unknown.

Draw *ca* to represent the reaction R_1 (= force CA) to a suitable scale in figure 8.24(*b*). Proceed round the space diagram of the joint in a

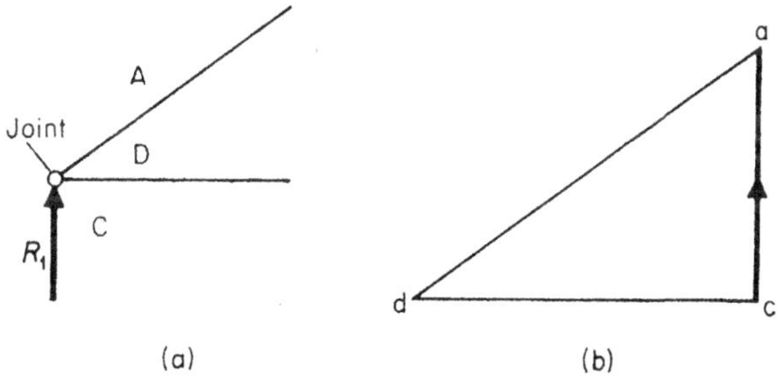

(*a*) Space diagram (*b*) Force diagram

Figure 8.24 The first joint—1st stage.

clockwise direction and draw lines parallel to the members AD and DC in turn, to obtain the force diagram *cad.*

The direction of the force CA determines the direction to follow round the force diagram, and so to add, if necessary, the arrow heads to forces *ad* and *dc* as shown in figure 8.25.

Having determined the directions of the arrow heads from the force diagram similar arrowheads should be marked at the appropriate joint on the space diagram as shown in figure 8.25(*b*).

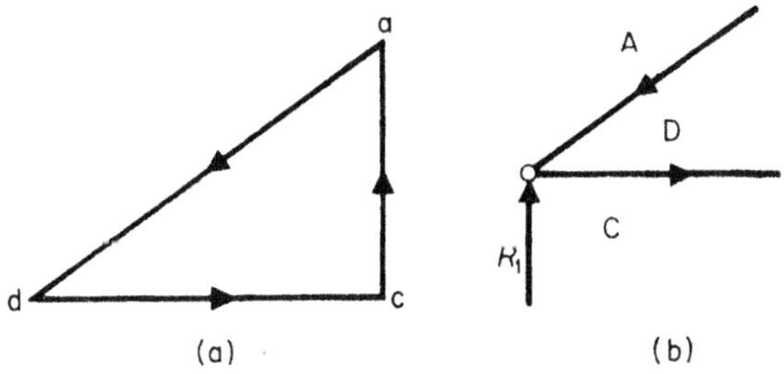

(*a*) The directions of the forces in the force diagram (*b*) The forces at the joint

Figure 8.25 The first joint—2nd stage.

But a 'pull' force at one end of a structural member indicates that the member is a tie and this force must be balanced by another 'pull' force at the opposite end of the member.

A second arrow head, opposing the first in direction, should therefore be added to member DC in the space diagram, as in figure 8.26(*a*). This member can now be recognized as ***a tie***.

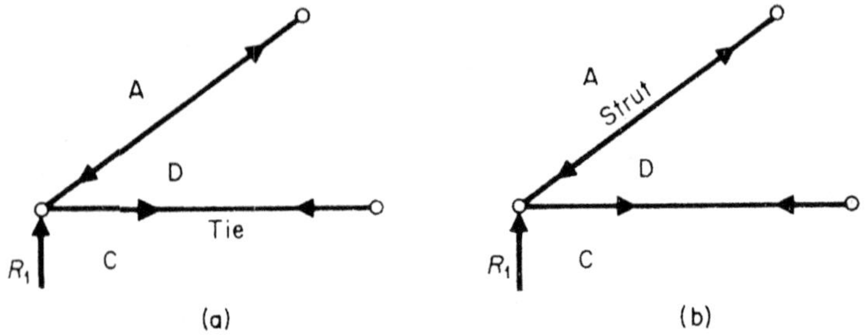

(*a*) The tie (*b*) The strut

Figure 8.26 The first joint—3rd stage.

Similarly, because a 'push' force at one end of a member indicates that the member is a strut another push force should be added to the other end of member AD to produce a state of equilibrium, as in figure 8.26(*b*). This member can now be recognized as *a strut*.

Now proceed to an adjacent joint, where at least one force is known and where there are not more than two unknown forces. Note that the direction of the force has been reversed in moving from a joint at one end of a structural member to a joint at the other end.

Consider next the joint carrying the vertical load *W*. Taking each load in turn round the space diagram, shown in figure 8.27(*a*), in a

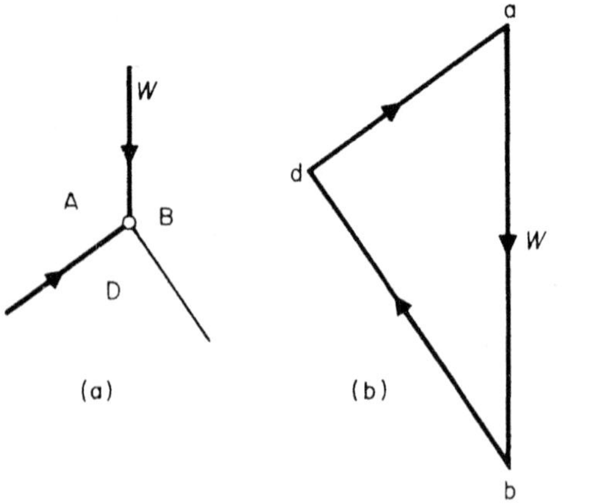

(*a*) Space diagram (*b*) Force diagram

Figure 8.27 The second joint.

clockwise direction, draw the force diagram to scale, and hence determine the magnitude and direction of the force in member BD.

The information regarding the direction of the forces should now be added to the space diagram shown in figure 8.28 indicating that member BD is a strut.

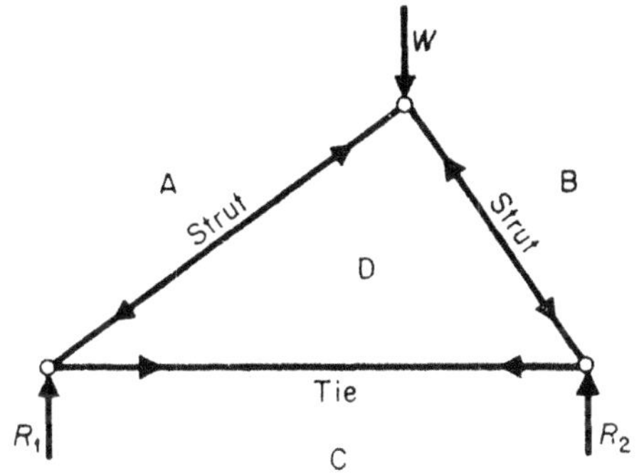

Figure 8.28 Forces in the members of the framework.

Clearly, having determined the magnitude of the forces in all the members there is no need to consider the right hand support and determine the forces from a force diagram at this joint. Its shape, however, is shown in figure 8.29.

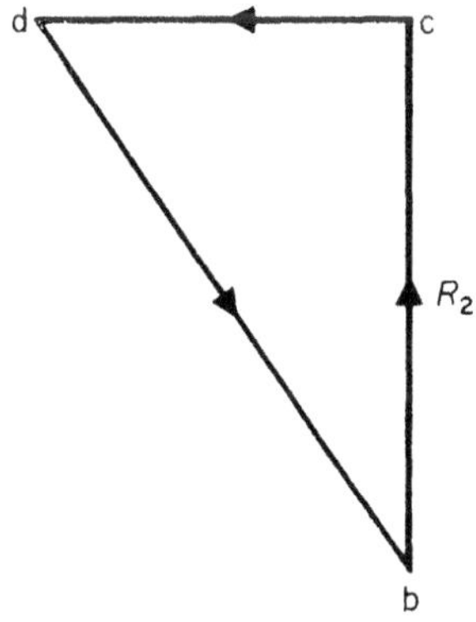

Figure 8.29 The third joint.

Now look at the separate force diagrams for the three joints; figures 8.25(*a*), 8.27(*b*) and 8.29. Note that the lines are duplicated, and, if the diagrams are all drawn to the same scale, they can be fitted together to form a composite force diagram as in figure 8.30. It is usual to draw such a diagram rather than a series of smaller diagrams for every joint.

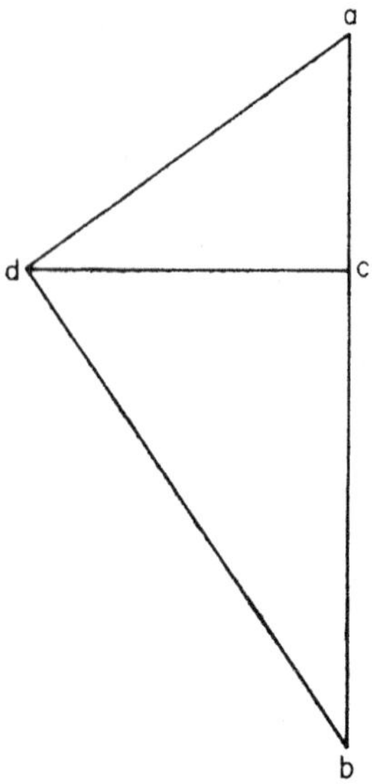

Figure 8.30 The composite force diagram.

If arrowheads were drawn on the composite diagram they would first be in one direction and then in the opposite direction on the same line when moving from one joint to the next. This could lead to confusion, and therefore *all arrowheads are omitted from the composite force diagram.* They are, however, placed on the space diagram as soon as the force diagram for a joint is completed.

Note that the force diagram for the external forces (the vertical line *acb*) forms a part of the composite diagram, and should be drawn first.

The results are best shown in tabular form, as follows:

Member	*Force lbf or tonf*	*Type*
CA		Reaction
AD		Strut
DC		Tie
BD		Strut
BC		Reaction

Example

A cantilever pin-jointed frame is shown in figure 8.31. Draw the force diagram and state the magnitude and type of force in the members

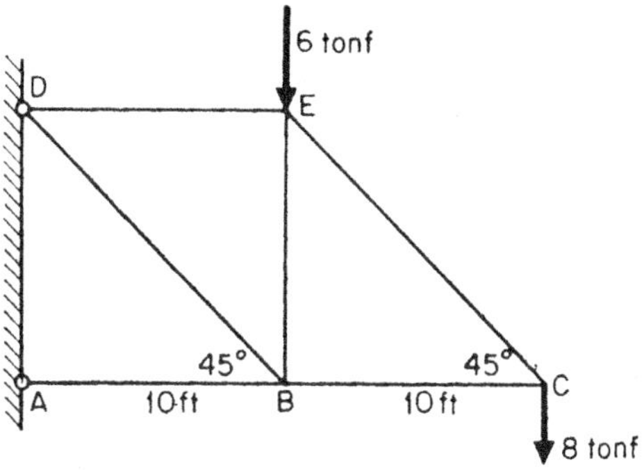

Figure 8.31

CE, EB and AB. Use the following scales: space diagram 1 in. to 5 ft; force diagram 1 in. to 4 tons. (U.L.C.I.)

Solution

First of all, letter the spaces in accordance with Bow's Notation. Without finding the reaction forces at A and D, the starting point for drawing the space diagram must be joint C; this is the only one at which there are only two members and therefore two unknown forces. This is an important consideration which the student should appreciate.

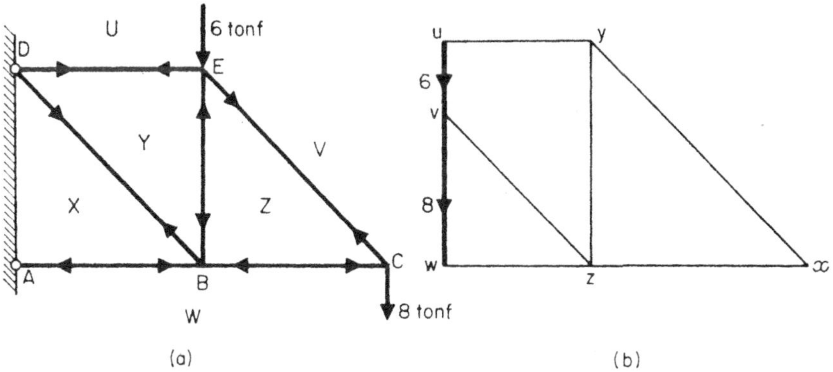

(*a*) Space diagram (*b*) Force diagram

Figure 8.32

vw is drawn to represent the 8 ton load. By drawing lines parallel to WZ and ZV through *w* and *v* respectively the point *z* is determined by intersection and hence the triangle of forces *vwz* is completed. The senses of the vectors, dictated by the sense of the load vector *vw* and continuing around the triangle, are then transferred to the space diagram at joint C. The arrows are reversed at joints E and B.

Joint E is now considered—there are still three unknowns at joint B.

uv, the 6 ton load vector is drawn, *vz* is known in accordance with the reverse arrow at E; lines parallel to ZY and YU drawn through *z* and *u* respectively enable the point *y* to be located by intersection. The polygon of forces *uvzy* is now completed The senses of the vectors can be determined by continuing around the figure in that same order which was fixed by the sense of the 6 ton load, and these again transferred as arrows to the appropriate members at joint E. The arrows are reversed at joints D and B.

Finally, joint B is considered. Vectors *yz* and *zw* are known, and by drawing lines through *w* and *z* parallel to WX and XY respectively the point of intersection, *x* is determined and the polygon is closed. Again sense arrows are transferred to joint B in the space diagram and reversed at joints A and D.

The forces required may be determined by scaling from the force diagrams; the answers are given in tabular form below.

Member	*Force vector*	*Force-tons*	*Type*
CE	*vz*	11·3	Tie
EB	*yz*	14·0	Strut
AB	*wx*	22·0	Strut

A further example, involving a nine-member framework, will now be worked but the detailed, step-by-step procedure will not be given in this case. The student is recommended to follow the reasoning independently and to draw out the space and force diagrams to the scales given in the question.

Example

A simply supported pin-jointed frame is shown in figure 8.33. Draw the force diagram and state the magnitude and type of the forces in the members PS, SW and RW. Use the following scales; length 1 in. to 5 ft; force diagram, 1 in. to 2 ton. (U.L.C.I.)

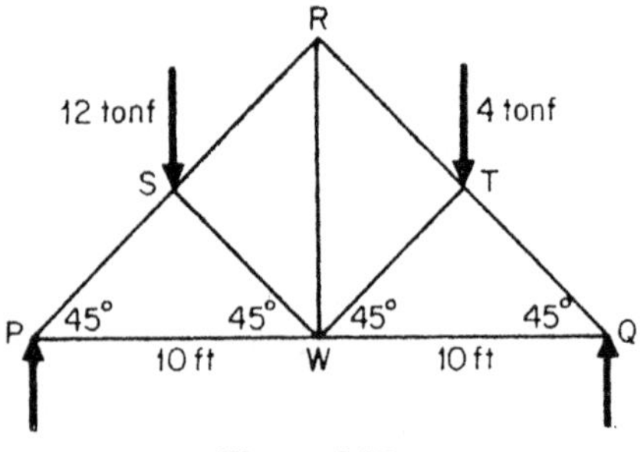

Figure 8.33

Solution

To find the support reaction. Take moments about P,

$$\begin{aligned}(R_Q \times 20) &= (12 \times 5) + (4 \times 15)\\ &= 120 \text{ ton}f\text{-ft}\\ R_Q &= 6 \text{ tonf}\end{aligned}$$

Take moments about Q,

$$\begin{aligned}(R_P \times 20) &= (12 \times 15) + (4 \times 5)\\ &= 200 \text{ ton}f\text{-ft}\\ R_P &= 10 \text{ tonf}\end{aligned}$$

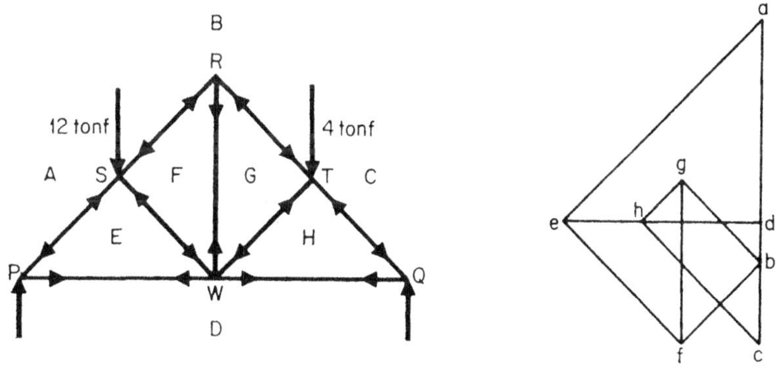

(a) (b)

(*a*) Space diagram (*b*) Force diagram

Figure 8.34

To construct the force diagram start at joint P and continue at joints S, R and T; it will be found that diagrams for joints Q and W follow automatically.

Results	*Member*	*Force vector*	*Force-tons*	*Type*
	PS	*ae*	14·14	Strut
	SW	*ef*	8·50	Strut
	RW	*fg*	8·0	Tie

Summary

The general conditions of equilibrium of a body under the actions of forces

(*a*) There must be no tendency for the body to be displaced in any direction, i.e. there must be no resultant of the force system.

(*b*) There must be no tendency of the body to rotate; i.e. the algebraic sum of the moments of the forces about any point in their plane must be zero.

Reactions of surfaces

At smooth surfaces the surface reaction is always in a direction normal to the surface.

At rough surfaces, where the action of an inclined force is tending to cause slip of a body on the surface, the surface reaction is not normal to the plane.

Solution of problems of ladders and trapdoors in equilibrium is facilitated if it is remembered that if a body is in equilibrium under the action of three non-parallel forces, those forces must pass through the same point.

Bow's notation

This is a system of lettering of space and force diagrams, which is much used in the solution of frameworks. The *spaces* between the forces marked on the space diagram are lettered with capital letters. Each force is then known and denoted by the letters on either side of it.

The vector diagram is lettered with the same letters, but in this case, the lines representing the forces are lettered end-to-end.

Forces in the members of simple framed structures

Frameworks may be solved by the use of Bow's notation. Sometimes it is first necessary to take moments round one support to find the reaction at the other.

The starting point for the graphical solution must be a joint at which not more than two unknown forces act. Each joint is separately solved, the forces due to the members being discovered by drawing a vector diagram. The directions of these forces are denoted by arrow-heads adjacent to the joint on the space diagram.

The solved forces on the space diagram are then reversed to act at the member ends remote from those which were the subject of the previous force diagram, in order to enable the solution of the other joints. These are then analysed by drawing the next part of the continuous vector diagram. During each analysis it is necessary to work in the same direction (e.g. clockwise) round all the joints being solved.

Exercises

1. A uniform ladder 16 ft long weighs 40 lbf. One end rests against a smooth vertical wall and the other stands on a rough horizontal surface at a point 4 ft from the foot of the wall. Determine the magnitude and direction of the reactions of the vertical and horizontal surfaces, and state what assumptions you make.

2. A uniform ladder is 8 metres long and mass 24 kg. It rests against a smooth vertical wall and upon a rough horizontal surface at a point 2·24 metres from the foot of the wall. Determine the reactions of the surfaces at the two ends of the ladder.

3. A trapdoor 5 ft long, weighs 40 lbf, and is fastened to a vertical wall by a hinge. At the other end is a rope which is 6 ft long and secured to a point in the wall 8 ft above the hinge. Determine

(*a*) the tension in the rope, and

(*b*) the magnitude and direction of the hinge reaction.

4. A uniform bar 3 metres long and of mass 120 kg is hinged at one end to a vertical wall. The other end is held by a horizontal rope to a point on the wall 1·2 metres above the hinge, find

(*a*) the tension in the rope, and

(*b*) the magnitude and direction of the hinge reaction.

5. A uniform beam 7 ft long is hinged to a wall and held in a horizontal position by a rope attached to its free end, and making an angle of 40° with the beam. If the weight of the beam is 100 lbf, find

(*a*) the tension in the rope, and

(*b*) the magnitude and direction of the hinge reaction.

6. A symmetrical framework is simply supported at its ends loaded as

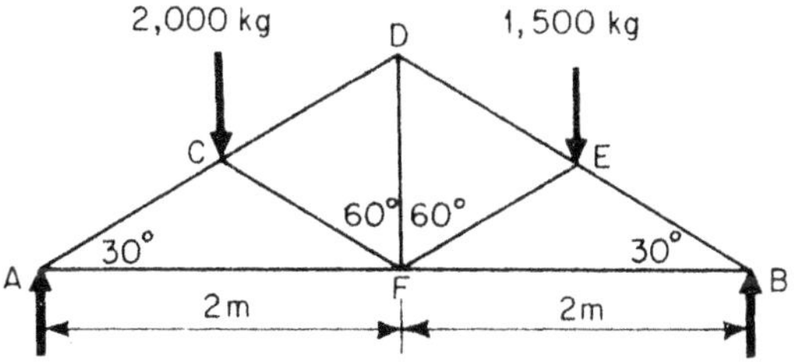

shown. Determine the reactions at the supports and the magnitude and type of loading in each member. Tabulate your results.

7. A pin-jointed structure is simply supported at its ends and carries

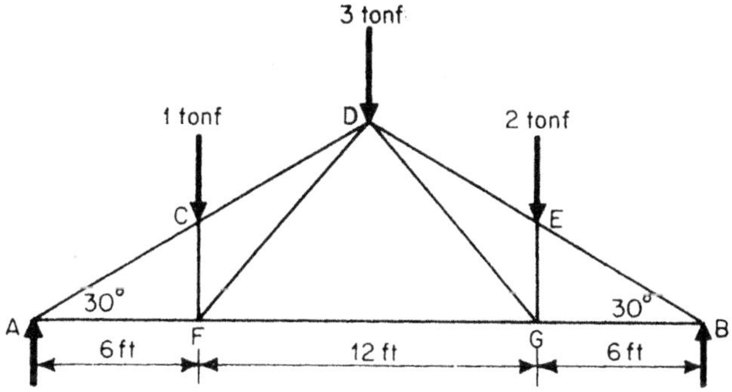

vertical loads as shown. Find the magnitude of the support reactions, and also the magnitude and type of force in each member of the structure.

8. Find the forces in each of the members of the framework shown when

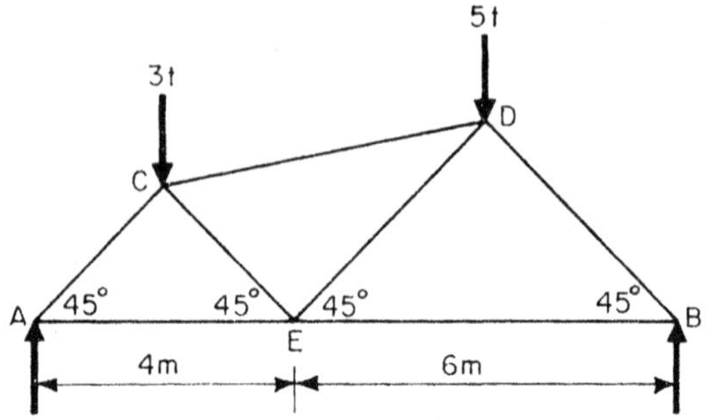

it is supported at its ends and carries vertical loads. Tabulate the results showing which members are in tension and which are in compression. Loads are in tonnes mass.

9. Find the force in each of the members of the framework shown,

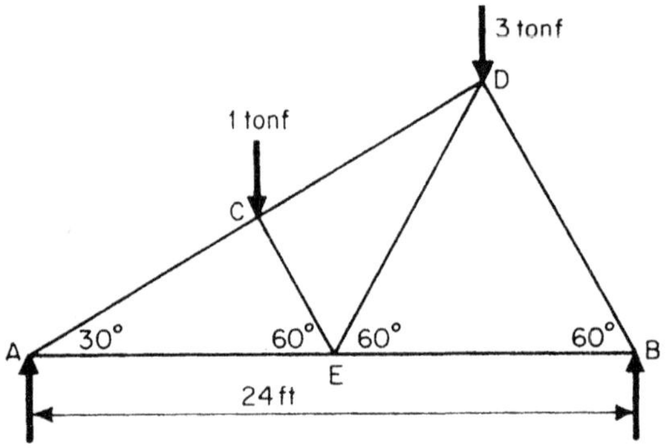

when it is simply supported at its ends. State which members are in tension and which in compression.

10. The figure illustrates part of a bridge girder. For the loading shown

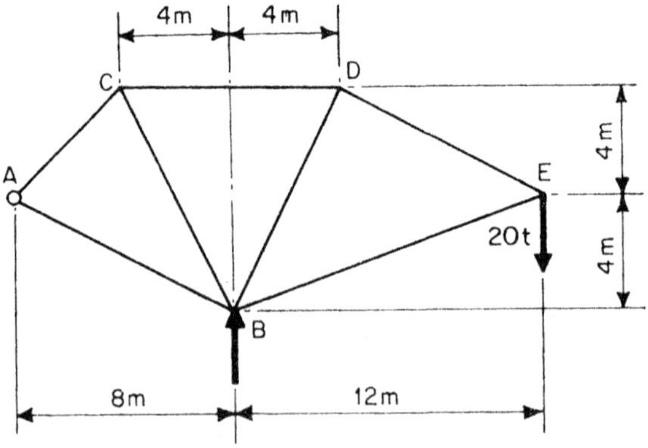

draw a scaled force diagram and hence determine the forces in all seven members of the frame.

9

Mechanical properties of engineering materials

Chapter 8 deals with the determination of the forces acting in the members of a structure as a result of externally applied forces. The mechanical properties of those materials which are commonly used in engineering need to be understood to determine the size and shape of each member to carry its respective load. First it is necessary to define some of these properties.

Elasticity
If a material, deformed by the application of a force, returns to its original dimensions when the force is withdrawn, the material is said to be elastic. Elasticity is the property of a material which permits it to change its shape temporarily and return to its original shape when an external force no longer acts on it.

Plasticity
A material which is deformed easily without fracture by the application of a force, and does not return to its original dimensions when the force is withdrawn, is said to be plastic. A material which exhibits elastic properties under a given range of loading may become plastic when subjected to greater loads. Under the action of heat most materials become plastic.

Strength
The strength of a material is its ability to resist the application of an external force without fracture occurring. If the internal resistance of the material, or the cohesion of its molecules, is insufficient to resist the external force the material fractures.

Ductility
In the process of wire-drawing it is necessary to draw out, or elongate, the material in a longitudinal direction, and at the same time to reduce its cross-sectional area. A material which can be drawn out easily is said to be ductile. Such a material possesses considerable plasticity and also strength. During the drawing out process the material becomes progressively stronger, and hence more resistant to deformation.

Malleability
Some materials can be changed in shape and size by hot-rolling, forging, drop-stamping and hammering. If deformation occurs without fracture the material is said to be malleable. Such a material must possess considerable plasticity, but not necessarily great strength.

Brittleness
If a material fractures under the influence of an external force which is applied slowly, and does not in the process deform appreciably the material is said to be brittle. Such a material fractures easily when subjected to shock loading. Brittleness is the converse of ductility.

Toughness
The property of a material which has a high resistance to impact loading.

Hardness
This is a measure of the ability of a material to withstand abrasion, scratching and indentation by other materials.

All the above properties are relative. Some common basis for comparison is required to determine the *degree* of hardness, or toughness of a material. Consequently standard methods of testing materials have been devised, and suitable machines have been designed and developed for use in a laboratory.

Direct force (symbol P)
Members of a structure may be loaded axially either by applying a 'pull' force or a 'push' force. Such forces are known as direct forces. They are the simplest forms of loading.

A direct force is usually determined in pounds force (abbreviated to lbf) or tons force (abbreviated to tonf). The corresponding metric unit of force, or static load, is the kilogram force (abbreviated kgf).

Direct stress (symbol σ—Greek letter sigma)
An external direct force is resisted by internal forces of cohesion. That is, the molecules in the material tend to stick together and resist separation. The value of the direct force divided by the cross sectional area of the member gives the average stress or intensity of loading, over the

area. It is usual to assume that the intensity of loading is uniform over the whole section, that is

$$\text{Average stress} = \frac{\text{Direct force}}{\text{Cross-sectional area}}$$

or,

$$\sigma = \frac{P}{A}$$

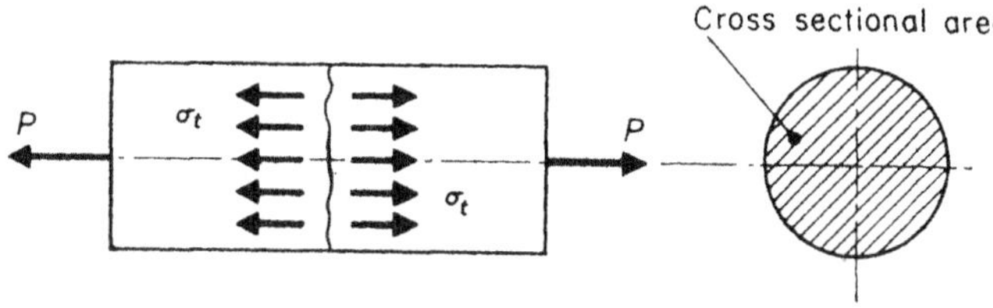

Figure 9.1 A 'pull,' or tensile force.

If the external force P is a tensile force, figure 9.1, the average stress set up on any cross-sectional area is known as the tensile stress, and has the symbol σ_t.

If the external force P is a compressive force, figure 9.2, the average stress set up on any cross-sectional area is known as the compressive stress, and has the symbol σ_c.

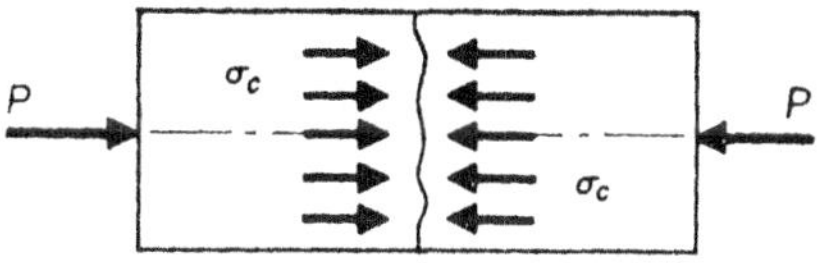

Figure 9.2 A 'push', or compressive force.

A direct stress is usually determined in lbf/in^2 or tonf/in^2. The corresponding metric unit is kilogram force per square metre—kgf/m^2. The SI unit is the newton per square metre—N/m^2.

Direct strain (symbol ε—Greek letter epsilon)

When a direct force is applied to a structural member it tends to lengthen if the force is tensile, or to shorten if the force is compressive.

Strain is a measure of the deformation, and direct strain is the deformation determined along the line of action of the direct force.

$$\text{Direct or linear strain} = \frac{\text{Change in length}}{\text{Original length of member}}$$

or,

$$\varepsilon = \frac{\delta l}{l}$$

Note that strain, being a ratio of two lengths, has no units, and is therefore just a number.

Tensile test of a mild steel specimen

If a length of mild steel bar is subjected to an axial tensile load the extension of the specimen is, at first, very small and would be undetectable by direct measurement, such as with a scale. It is necessary, therefore, to make use of an instrument known as an extensometer. This instrument is clamped over a known length of the specimen, and the elongation of that length is magnified so that it can be determined easily.

As the tensile load increases so the extension increases until a point is reached where the extension increases rapidly and becomes measurable by dividers and a scale.

Later, there is a noticeable reduction, or waisting, of the cross-sectional area of the specimen over a limited length, as shown in

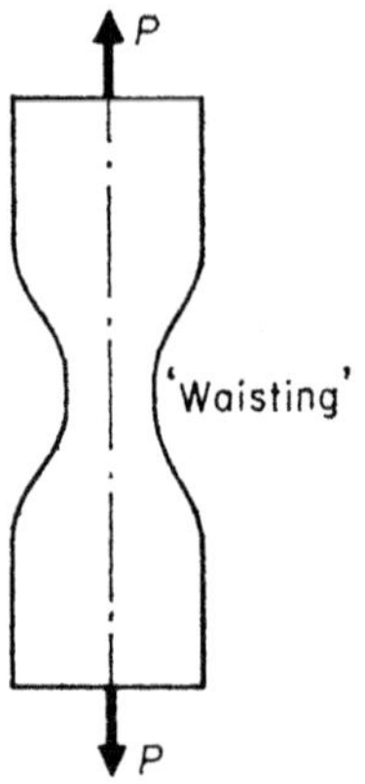

Figure 9.3 'Waisting' of a tensile specimen under load.

figure 9.3 and extension of this reduced portion takes place very rapidly. This 'waisting' is a typical phase in the stretching of all ductile materials.

Finally, the specimen fractures at the smallest section.

If the values of load and corresponding extension are plotted, 'load' usually being plotted vertically and 'extension' horizontally, the graph is of the form shown in figure 9.4.

The first portion of the graph is found to be a straight line, and experiments show that if the tensile load is removed at any time during the loading O to A the piece of steel will return to its original dimensions. The material, therefore, is elastic in this region. The point A, at the end of the straight line portion, is called *the limit of proportionality*.

If the load is removed when the material has stretched to a point, beyond A on the graph, the specimen will not return to its original dimensions even after a long period of time. It will have a permanent

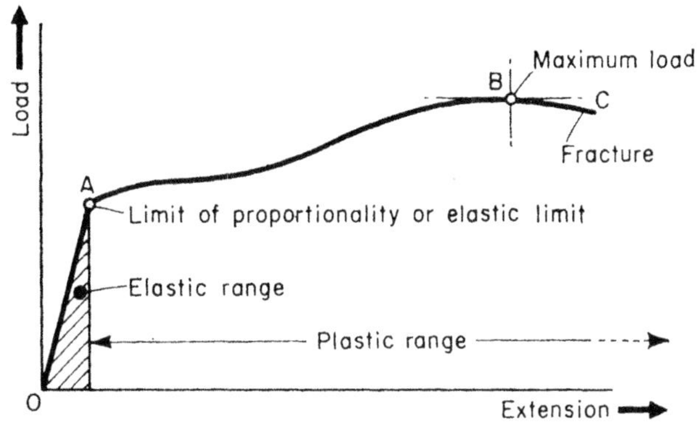

Figure 9.4 Load—Extension graph of a tensile test on a mild steel specimen.

set. In this region, from A to the fracture at C, the extension is very large for small increases in the load. Consequently the material is said to be plastic.

Note that this term is not connected with the word 'plastics'.

If the extension is maintained at a constant rate during the tensile test, it is found that the load has to be reduced when waisting begins. Consequently the maximum load occurs before fracture, at a point B.

Hooke's law

A straight line portion of the load-extension graph occurs with most metals when subjected to a direct load. Hooke's law states that in any elastic material there is a linear relationship between applied stress and strain, provided the limit of proportionality is not exceeded.

Young's Modulus, or the Modulus of Elasticity (symbol E)
It is usual in a tensile test to determine the stress in the material for each value of the applied load. The value of the stress is obtained by dividing the load by the *original* cross-sectional area: that is, the cross-sectional area determined at the beginning of the test.

Hence, $$\text{Stress} = \frac{\text{Applied load}}{\text{Original cross-sectional area}}$$

or, $$\sigma = \frac{P}{A}$$

and $$\text{Strain} = \frac{\text{Change in length}}{\text{Original length}}$$

or, $$\varepsilon = \frac{\delta l}{l}$$

Because both A and l are constants

$$\sigma = \text{Constant} \times P$$

and $$\varepsilon = \text{Constant} \times \delta l$$

If, therefore, a second graph is plotted with stress vertical and strain horizontal, it will be similar to the one shown in figure 9.4 but to a different scale, and there will be, as before, a straight line relationship over the elastic range.

The ratio $\frac{\text{stress}}{\text{strain}}$ corresponds to the slope of this straight line, and its value must be constant over this limited range. For any given material the value will be sensibly the same, but each different material will have its own particular value.

This constant is called Young's Modulus, or the Modulus of Elasticity, and is denoted by the symbol E.

Thus $E = \frac{\sigma}{\varepsilon}$ within the limit of proportionality.

Because strain is a ratio of two lengths, and hence a number, the units of E must be the same as those of σ, namely, lbf/in^2 or tonf/in^2 or, in metric units, kN/m^2 or, (SI), N/m^2.

There are, of course, many types of most materials in general use, and therefore the values of E, obtained from experiments, are only approximate.

Average values of the modulus for a few metals are given in the following table:

Material	*Modulus of Elasticity (E)*		
	lbf/in²	*tonf/in²*	*N/m²*
Mild Steel	30×10^6	13,400	$20{\cdot}6 \times 10^{10}$
Copper	17×10^6	7,600	$11{\cdot}7 \times 10^{10}$
Aluminium alloys	10×10^6	4,450	$6{\cdot}8 \times 10^{10}$
Brass	12×10^6	5,350	$8{\cdot}8 \times 10^{10}$

Note: the newton, the standard SI unit of force, is also the standard for derived units of pressure, stress, and the various moduli.

Ultimate tensile stress (abbreviated to U.T.S.)

The ultimate tensile stress is the ratio of the maximum load applied in a tensile test of the material to its original cross-sectional area.

$$\text{U.T.S.} = \frac{\text{Maximum load}}{\text{Original cross-sectional area}}$$

This stress is used as a basis for the design of a structural member.

In B.S. 1991, the Ultimate Tensile Stress is called the Tensile Strength.

Maximum allowable stress

Although the U.T.S. is used as a *basis* for design, such a high value could never be permitted to occur in practice. Instead a much lower value is adopted, called the *maximum allowable stress*, or the *safe working stress*; this is always within the limit of proportionality of the material used, so that permanent distortion is avoided.

The choice of a safe working stress is derived from experiment and experience. A structure must be sufficiently strong to withstand all foreseeable conditions of loading and yet must be economical in the use of materials.

Factor of Safety

The Ratio $= \dfrac{\text{Ultimate tensile stress}}{\text{Maximum allowable stress}}$ is a number, and is known as the factor of safety. The value of this number always exceeds unity, and depends on the conditions which the designer has to meet. Many of these conditions are too complex to discuss here.

Example

(*a*) Define (i) *Modulus of elasticity*, and (ii) *Yield point*.

(*b*) A mild steel specimen of 0·564 in. diameter was found to have an extension of 0·008 in. on an 8 in. gauge length when a load of $3\frac{1}{2}$ tonf was applied. Determine Young's modulus for the material assuming the load to lie within the limit of proportionality. If a load of 7 tonf produced an extension of 0·036 in. on the 8 in. gauge length, determine whether the stress would still be within the limit of proportionality. (N.C.T.E.C.)

Solution

(*a*) (i) The modulus of elasticity is defined earlier in this chapter.

(ii) The yield point is the load or stress at which extension of a specimen in a tensile test becomes measurable by means of a scale. It is not of any practical significance.

(*b*) Diameter, d, of specimen $= 0{\cdot}564$ in.

Cross-sectional area $A = 0{\cdot}7854 \times (0{\cdot}564)^2$ in^2

$= 0{\cdot}250$ in^2

$$\text{Strain} = \frac{\delta l}{l} = \frac{0{\cdot}008}{8} = 0{\cdot}001$$

$$\text{Stress} = \frac{W}{A} = \frac{3\frac{1}{2}}{0{\cdot}250} = 14 \text{ tonf/in}^2$$

$$\text{Young's modulus, } E = \frac{\text{Stress}}{\text{Strain}} = \frac{14}{0{\cdot}001}$$

$$= 14{,}000 \text{ tonf/in}^2$$

If stress were still within limit of proportionality, extension would be proportional to load.

$$\text{Proportional extension} = \frac{7}{3\frac{1}{2}} \times 0{\cdot}008 = 0{\cdot}016 \text{ in.}$$

Since extension of 0·036 in. is more than twice this figure, the stress is outside the limit of proportionality.

Example

A hollow cast iron section is 25 cm outside diameter, the metal being 2·5 cm in thickness.

If the permissible compressive stress in the cast iron is $82{\cdot}4 \times 10^6$ N/m^2 calculate the maximum load which could be safely carried by a short column made up from this section.

Given E for cast iron is $10{\cdot}79 \times 10^{10}$ N/m^2, what change in length would result from the application of the maximum permissible load to a column 4·5 metres high?

Solution

$$\text{Cross-sectional area of metal} = \frac{\pi}{4}(25^2 - 20^2) \text{ cm}^2$$

$$= 0{\cdot}7854 \times 5 \times 45 \text{ cm}^2$$

$$= 176{\cdot}6 \text{ cm}^2$$

$$\text{Maximum permissible load} = 82{\cdot}4 \times 10^6 \text{ N/m}^2 \times \frac{176{\cdot}6}{10^4} \text{ m}^2$$

$$= 1455{\cdot}18 \text{ kN}$$

$$\text{Strain} = \frac{\text{Stress}}{E}$$

$$= \frac{82{\cdot}4 \times 10^6 \text{ N/m}^2}{10{\cdot}79 \times 10^{10} \text{ N/m}^2}$$

$$= 0{\cdot}763 \times 10^{-3}$$

$$\text{Change in length of column} = 4{\cdot}5\ \text{m} \times 0{\cdot}763 \times 10^{-3}$$
$$= 3{\cdot}43 \times 10^{-3}\ \text{m}$$
$$= 0{\cdot}343\ \text{cm}$$

(*Note:* The SI unit of stress is the newton per square metre (N/m^2).)

Note that, $1\ kN/m^2 = 1000\ N/m^2$

Example

State 'Hooke's Law'.

Define clearly (1) stress, (2) strain.

A tensile test of a structural mild steel showed a maximum load of 7·5 tonf on a specimen of 0·564 inches diameter. Calculate in inches the extension which would occur, under working load, on a tie rod of this material 12 ft long if a design factor of safety of 3 is used and Young's Modulus for the material is 13,000 tonf/in².

$$\text{Factor of safety} = \frac{\text{Ultimate Stress}}{\text{Working stress}}$$ (N.C.T.E.C.)

Solution

$$\text{Cross-sectional area} = 0{\cdot}7854 \times (0{\cdot}564)^2\ \text{in}^2$$
$$= 0{\cdot}250\ \text{in}^2$$
$$\text{Ultimate tensile stress} = 7{\cdot}5\ \text{tonf} \div 0{\cdot}250\ \text{in}^2$$
$$= 30\ \text{tonf/in}^2$$
$$\text{Working stress} = \text{U.T.S.} \div \text{Factor of safety}$$
$$= \frac{30}{3}\ \text{tonf/in}^2$$
$$= 10\ \text{tonf/in}^2$$
$$\text{Extension} = \text{Strain} \times \text{Original length}$$
$$= \frac{\text{Stress}}{E} \times \text{Original length}$$
$$= \frac{10\ \text{tonf/in}^2 \times 144\ \text{in.}}{13{,}000\ \text{tonf/in}^2}$$
$$= 0{\cdot}111\ \text{in.}$$

Shear force (symbol Q)

Referring again to figure 9.1 a direct force of P on one side of the selected cross-section is opposed by an equal direct force of P on the other side to maintain equilibrium, both forces being axial. Sometimes the applied force is opposed by an equal force acting in a direction parallel to, but not coincident with it. A riveted joint is an example of this type of loading. Consider a joint formed by two pieces of overlapping metal, held together by a single rivet, as in figure 9.5.

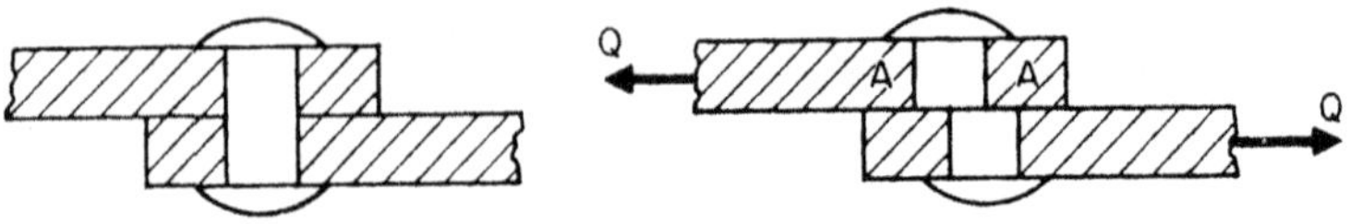

Figure 9.5 A riveted joint.

The effect of applying a force Q to each of the pieces of metal is a tendency for the top half of the rivet to slide over the bottom half along the line AA. The rivet is said to be in shear, the shear force being denoted by the symbol Q.

Shear stress (symbol τ—Greek letter, tau)
The shear force induces stresses in the rivet along line AA and *parallel to the direction of the applied force.*

It is assumed that the shear stress will be uniform across the area of the rivet in shear, and therefore

$$\text{Shear stress} = \frac{\text{Shear force}}{\text{Cross-sectional area of rivet}}$$

Example

A cast iron gear wheel is 7·6 cm wide. The mean thickness of each tooth at the pitch line is 1·9 cm.

What is the average shear stress in the tooth when a force of 2,000 N, directed tangentially along the pitch line, acts on the tooth?

Solution

$$\text{Area of tooth resisting shear} = 7{\cdot}6 \times 1{\cdot}9 \text{ cm}^2$$
$$= 14{\cdot}43 \text{ cm}^2$$
$$\text{Average shear stress} = \frac{\text{Shear force}}{\text{Area resisting shear}}$$
$$= \frac{2{,}000 \text{ N}}{14{\cdot}43 \text{ cm}^2}$$
$$= 138{\cdot}5 \text{ N/cm}^2$$
$$= 13{\cdot}85 \times 10^5 \text{ N/m}^2$$

or,
$$= 13{\cdot}85 \times 10^2 \text{ kN/m}^2$$

Punching a hole in a plate

To form a hole in a plate by means of a cylindrical punch, an axial force P is applied to the punch as in figure 9.6.

As the punch proceeds downwards it will force a cylindrical piece of metal out of the plate.

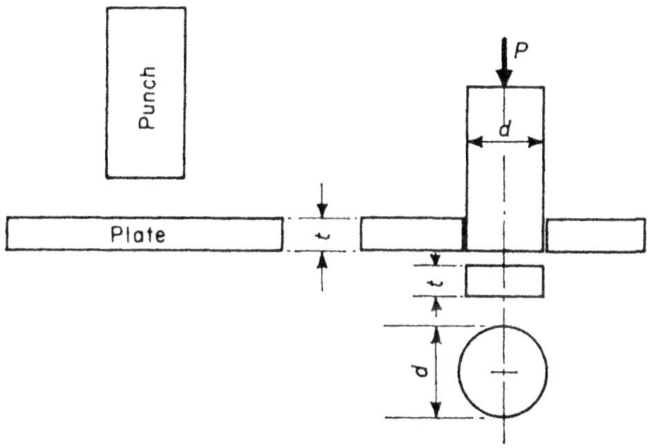

Figure 9.6 Punching a hole in a plate.

The area of the plate resisting the action of the punch is equal to the curved surface area of the cylindrical disc which is forced out, that is its circumference multiplied by the thickness of the plate.

$$\text{Area resisting shear} = \pi dt$$

If P is the steady force applied to the punch axially, and τ is the maximum shear stress which the plate can withstand before shear takes place, then, because

$$\text{Shear force} = \text{Shear stress} \times \text{Area resisting shear}$$
$$Q = \tau \times \pi d \times t$$

The compressive stress (σ_c) in the punch is determined from the expression

$$\text{Applied force} = \text{Compressive stress} \times \text{Cross-sectional area of punch}$$

or, $$P = \sigma_c \times \frac{\pi}{4} d^2$$

The minimum force applied to the punch must be equal to the force exerted by the plate in resisting shear. That is

$$P = Q$$

or, $$\sigma_c \times \frac{\pi}{4} d^2 = \tau \times \pi dt$$

Example

Calculate the axial force required to punch a hole $\frac{1}{2}$ in. diameter in a plate $\frac{3}{8}$ in. thick if the shear strength of the material is 30 tonf/in^2.

Solution

$$\begin{aligned} \text{Area resisting shear} &= \pi d \times t \\ &= \pi \times \tfrac{1}{2} \times \tfrac{3}{8}\ \text{in}^2 \\ &= 0{\cdot}589\ \text{in}^2 \\ \text{Force to punch hole} &= \tau \times \text{Area resisting shear} \\ &= 30\ \text{tonf/in}^2 \times 0{\cdot}589\ \text{in}^2 \\ &= 17{\cdot}67\ \text{tonf} \end{aligned}$$

Summary

The mechanical properties of materials used in engineering
These include elasticity, plasticity, strength, ductility, malleability, toughness, brittleness, and hardness.

Force and stress
Direct forces are those which act normal to the section under consideration. Stress is defined as force per unit area

or,
$$\text{Stress} = \frac{\text{Force}}{\text{Area}}$$

Direct stresses are either tensile or compressive.

Strain
Strain is an index of deformation. Direct strain corresponding to direct stress is defined as

$$\frac{\text{Change in length}}{\text{Original length}}$$

Tensile test on a mild steel specimen
The common indexes of material properties are usually derived from tensile tests. In these a test-piece of standard form is axially loaded in a tensile machine until fracture occurs.

(*a*) In the low-stress area of the test, the material stretches *elastically* until the elastic limit, or *limit of proportionality* is reached.

(*b*) After the limit of proportionality is passed, the plastic zone is entered; if the load is now removed, the material does not return to its original proportions.

Hooke's law

This states that in any elastic material loaded within the elastic range, there is a linear relationship between stress and strain.

Young's modulus

The ratio stress/strain of a material loaded within the elastic range is called 'Young's modulus of elasticity'. It is denoted by the symbol E.

Using the symbols of this chapter,

$$E = \frac{\sigma}{\varepsilon} \quad \text{or} \quad \frac{\frac{P}{A}}{\frac{\delta l}{l}} = \frac{Pl}{A\,.\,\delta l}$$

Ultimate tensile stress ('Tensile strength': B.S.:1991)

This index is given by the ratio

$$\frac{\text{Maximum load}}{\text{Original cross-sectional area}}$$

Factor of safety

Engineering components are never designed to be stressed to ultimate values in service. In practice the ultimate stress is divided by some factor to give the design value of working stress. This factor is called the 'factor of safety'.

Maximum allowable (working) stress × Factor of safety = Ultimate stress

Shear force and stress

Materials loaded so that adjacent sections tend to slide across each other are said to be 'in shear'. Such loading is called shear force. The value of the *shear stress* is derived by dividing this force by the area of metal under shear.

Exercises

1. A mild steel bar $\frac{1}{2}$ in. diameter and 10 in. long is subjected to an axial load of 2 tonf. If the modulus of elasticity of the material is 30×10^6 lbf/in^2, calculate (*a*) the average direct stress and (*b*) the extension of the bar.

2. An aluminium rod 0·90 m long, and 2·5 cm diameter has each extremity turned down in a lathe to 2·0 cm for a length of 0·15 m. The rod is then subjected to a tensile load of 78·5 kN. Calculate (*a*) the stress in each length, and (*b*) the total extension of the rod, given that Young's Modulus for aluminium is $6{\cdot}86 \times 10^{10}$ N/m^2.

3. Determine the load which will produce a stress of 10 tonf/in^2 in a piece of wire of $\frac{1}{8}$ in. diameter. If the length of the wire is 10 ft and Young's Modulus for the material is 13,000 tonf/in^2 what will be its extension under the load.

4. A tie rod 3 m long and 2·5 cm diameter stretches 0·18 cm when subjected to a tensile load of 61·8 kN. Calculate (*a*) the average stress (*b*) the strain and (*c*) the modulus of elasticity.

5. A cast iron column in the form of a tube has an outside diameter of 6 in. and an inside diameter of 5 in. It is subjected to a compressive load of 30 tonf. Calculate the average stress in the metal.

6. A mild steel specimen 1·0 cm diameter is stretched in a tensile testing machine to its limit of proportionality by a load of 22·56 kN. Calculate the maximum stress and the extension over a length of 5 cm if the modulus of elasticity is $21{\cdot}09 \times 10^{10}$ N/m^2.

7. A steel tie bar has a diameter of $1\frac{1}{2}$ in. Its ultimate tensile stress is 30 tonf/in^2. Find the safe load which the bar will carry if a factor of safety of 5 is to be used.

8. In a tensile test on a mild steel specimen the maximum load is found to be 34·33 kN. If the diameter of the test piece is 1·0 cm calculate the ultimate tensile stress.

Using a factor of safety of 3 calculate the working stress and the greatest allowable load on a 2·5 cm diameter bar of the same material.

9. Calculate the force required to punch a hole $\frac{7}{8}$ in. diameter in a plate $\frac{1}{2}$ in. thick if the shear strength of the material is 30 tonf/in^2. What would be the average compressive stress in the punch during the operation.

10. A steel column is made from a tube 15 cm outside diameter and 10 cm inside diameter. The column is 2 m long and carries a load of 100 t. If the modulus of elasticity of the steel is $19{\cdot}6 \times 10^{10}$ N/m^2, calculate the reduction in its length due to the load.

10

Beams—shearing force and bending moment

Apart from a passing reference to shearing force in Chapter 9 only the effects of direct load—either tensile or compressive—applied axially to a single member of a structure have been considered.

In many cases, however, a structural member is subjected to a transverse load; this tends to bend the member rather than lengthen or shorten it. A structural member which is subjected to transverse loads is called a *beam*.

In this course consideration will be restricted to beams which are horizontal and those in which the transverse loads are acting vertically downwards.

If a beam is to remain in equilibrium—that is, it does not move under the action of the loads—there must be forces acting vertically upwards to balance the loads which are acting vertically downwards. These upward forces act at the supports on which the beam rests, and are called *support reactions*.

Types of beam

Simply supported beam

A beam which rests on two supports without any end restraint, as in figure 10.1(*a*) is said to be simply supported. The supports are often,

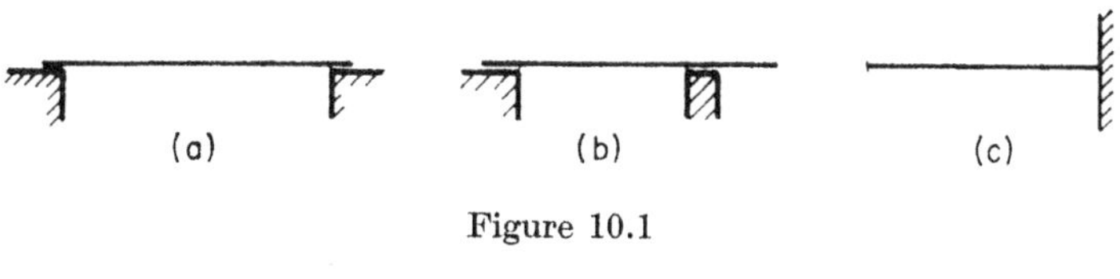

Figure 10.1

but not necessarily, situated at the ends. If a beam projects beyond its support it is called an overhanging beam. Figure 10.1(*b*).

Cantilever

If one end of a beam is fixed rigidly—that is, it is restrained from rotation—and the other end is free the beam is called a cantilever. Such a beam is supported at one end only—the fixed end, as shown in figure 10.1(*c*).

Types of load

Two types of load commonly occur. A load which does not extend over a measurable length of the beam is called a *concentrated load.* In calculations it is assumed that all the load is concentrated at a single point, and the space diagram is as shown in figures 10.2(*a*) and (*b*).

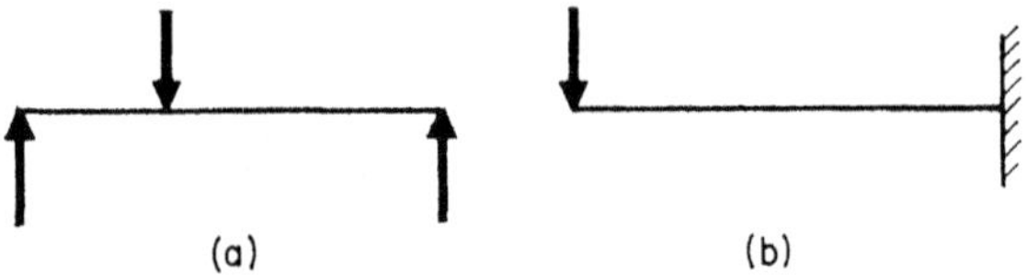

Figure 10.2

The cantilever diagram could, if necessary, be replaced by that shown in figure 10.3 indicating that at the fixed end, not only is there a vertical reaction, but also a clockwise turning moment resisting the tendency of the load at the free end to rotate the beam in an anti-clockwise direction.

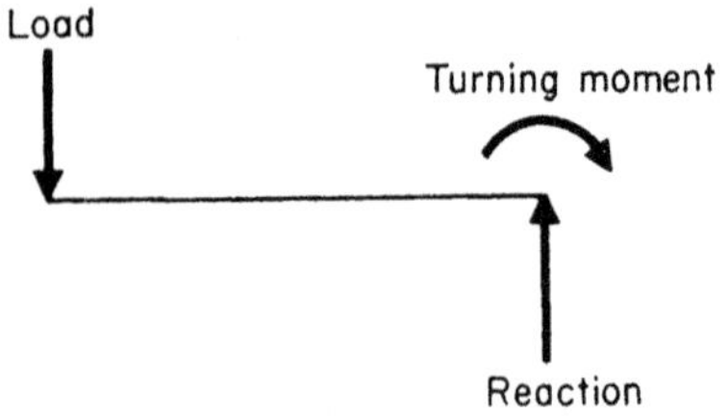

Figure 10.3 Forces acting on a cantilever.

From a practical point of view the load must be spread over some finite length, however small, but it is convenient to think in terms of point loading for all the applied loads (when concentrated) and also for the reactions at the supports.

If the load is spread over an appreciable length of the beam it is called a *distributed load.* Loads which exert an equal downward force for each unit length of the beam are known as uniformly distributed loads; they are represented diagrammatically as illustrated in figure 10.4.

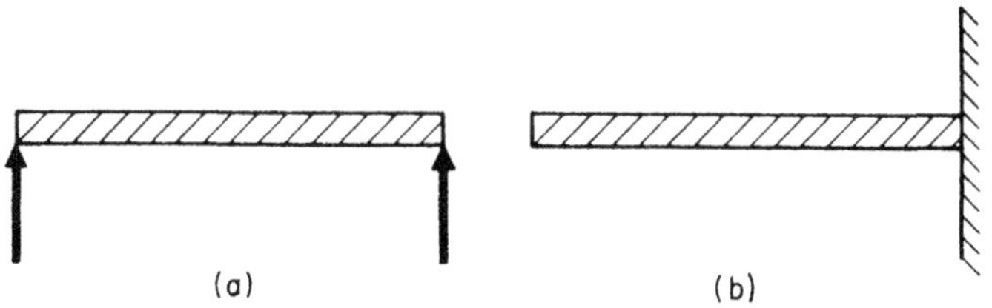

(*a*) Simply supported beam (*b*) Cantilever beam

Figure 10.4

Shearing force in a beam

A simply supported beam is loaded centrally with a concentrated load W and YY, figure 10.5(*a*), is a vertical section at a distance x measured from the left hand end of the beam.

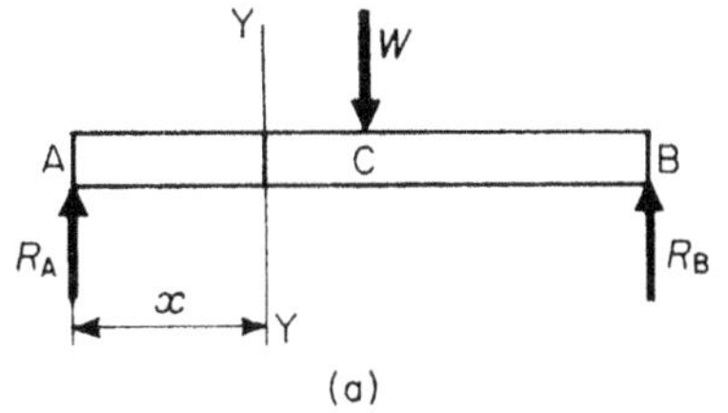

Figure 10.5(*a*) A simply supported beam, loaded centrally.

The vertical reaction R_A, at the left hand support, tends to cause the left hand portion of the beam to slide upwards relative to the right hand portion. This tendency is resisted by an internal force at the section perpendicular to the beam, shown diagrammatically in figure

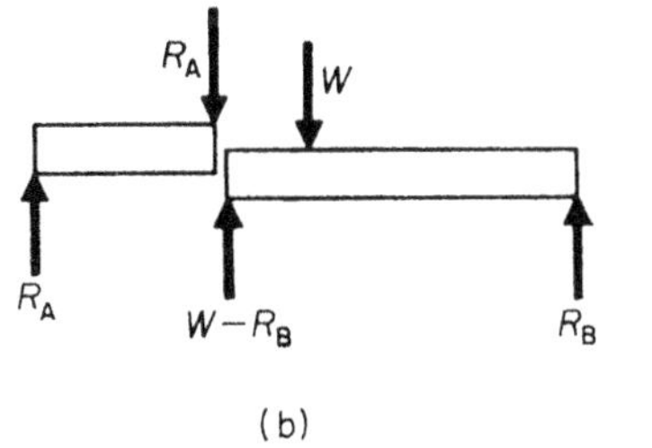

Figure 10.5(*b*) The shearing effect of a transverse load.

10.5(*b*). For equilibrium this force must be equal to R_A but act downwards. On the right hand portion of the beam two vertical forces are acting—W downwards and R_B upwards. By the same reasoning the force at section YY must be $W - R_B$, acting upwards. In the whole beam, these two internally induced forces must be equal and opposite to one another, i.e. $R_A = W - R_B$ and each is a measure of the shearing force at the section.

The magnitude of the shearing force at any section in a beam is equal to the algebraic sum—that is, taking upward and downward directions into consideration—of all the vertical external forces (including the reactions) acting on the beam *either* to the left *or* to the right of the section chosen. Thus if in figure 10.5 the vertical section YY is considered anywhere between the left hand extremity A and the centre C of the beam, the shearing force on the left hand portion of the beam cut off by YY is R_A upwards.

Similarly the shearing force on the right hand portion of the beam is $(W - R_B)$ downwards.

But, because the beam is in equilibrium, all vertical external forces acting on it must balance.

$$W = R_A + R_B$$

or,

$$R_A = W - R_B$$

Consequently, the shearing force R_A acting upwards on the left hand portion is equal to the shearing force $(W - R_B)$ acting downwards on the right hand portion of the beam.

Note that, in each case, the shearing force at the section YY is equal to the algebraic sum of all the vertical forces acting, either to the left or to the right of YY.

Convention

The sign of the shearing force is a matter of convention.

However, because it is customary to read from left to right, determine the distance x of each section from the *left* hand extremity of the beam.

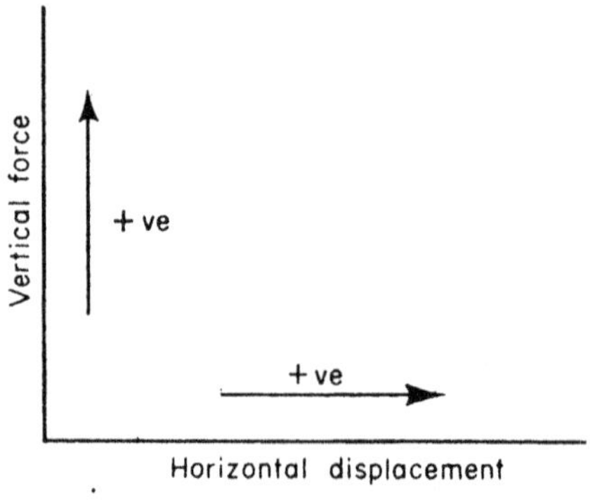

Figure 10.6 Sign of the shearing force.

Also, because when plotting a graph (*a*) the horizontal displacement is considered to be positive when it is measured to the right, and (*b*) the vertical displacement positive when it is measured upwards, this same convention is adopted for determining the signs of horizontal distance x and vertical forces, as in figure 10.6.

Thus in figure 10.5(*b*) x being measured from A towards B is POSITIVE, and R_A, being the upward shearing force on the left hand portion is also

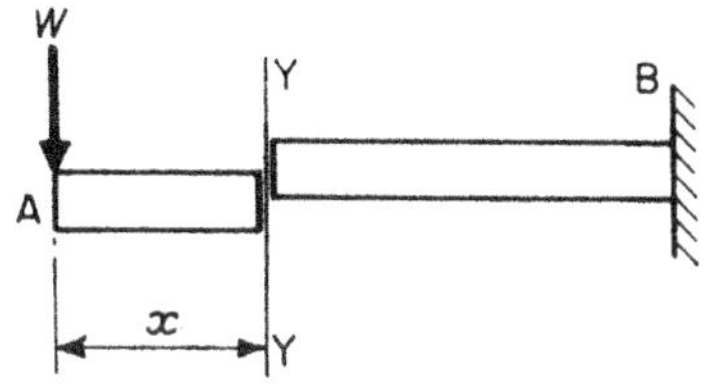

Figure 10.7 The shearing effect of a load on a cantilever.

POSITIVE, whereas in figure 10.7, x, being measured from A towards B is POSITIVE, but W, being the downward shearing force on the left hand portion is NEGATIVE.

Shearing force diagram

A shearing force diagram is a graph showing the variation of shearing force along the length of a beam. Such a diagram should be drawn immediately below the space diagram; it should be drawn to scale, and the scale stated in the correct units.

By reason of the convention to read values of x from left to right, and to regard upward forces as positive, the drawing of a shearing force diagram, which is a plot of shearing force values in a vertical direction and x values in a horizontal direction, will follow the normal procedure for plotting a graph.

1. Simply supported beam

(a) Concentrated load

A simply supported beam is one which rests on two supports. Suppose such a beam has a single load concentrated at its centre. In practice a load must be spread over a finite length, so that strictly the point load W at the centre, and the single point forces representing the reactions at each end should be replaced by narrowly spread loads as depicted in figure 10.8.

To the left of point A the shearing force is zero, whereas to the right of point A the shearing force is $+R_A$, and that over the support the shearing force varies from O to $+R_A$.

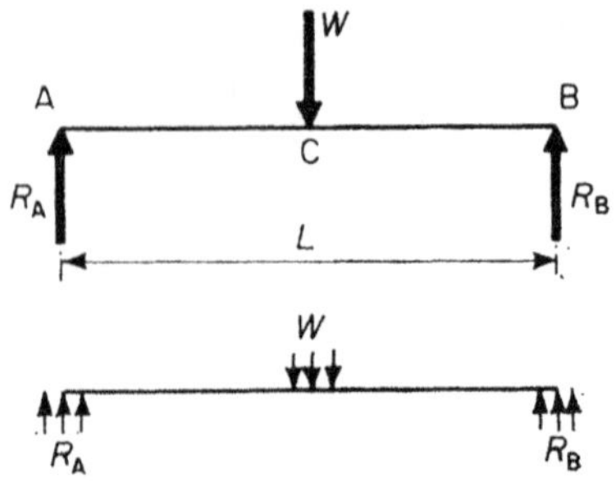

Figure 10.8 Concentrated loading.

When drawing a shearing force diagram, however, it is convenient to think in terms of point loading and not to be concerned with any gradual changes in the shearing force over a support or under a load.

In accordance with convention, start off with the left hand extremity of the beam and determine values of the shearing force as x increases from O to its maximum value of the span L.

At A where x equals O, the shearing force increases from O to $+R_A$. Because it is assumed that the forces are concentrated at single points select a convenient scale and plot R_A vertically above the datum line at O, as in figure 10.9. Between A and C the shearing force remains

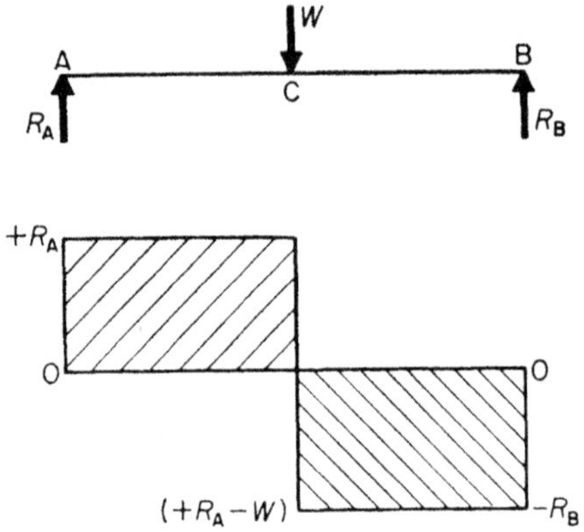

Figure 10.9 Shearing force diagram—simply supported beam with concentrated load.

constant at a value of $+R_A$. At C the shearing force changes suddenly, being reduced by W to value of $(R_A - W)$ and remains at this value from C to B. But

$$W = R_A + R_B$$

therefore $$R_A - W = -R_B$$

At B a positive shearing force is added, thus bringing the diagram back to the zero line at O.

Note that the value of the shearing force at any section is equal to the height of the diagram at that section. Consequently, between A and C the shearing force is $+R_A$, and, between C and B, is $-R_B$.

(*b*) *Distributed load*

Consider a simply supported beam with a uniformly distributed load spread over the whole span. Let the span be L, and let the load per unit length be w. This expression means that there is an evenly spread load of w over each unit length of the beam. In a practical example the units may be ton and foot or, in metric, kilogram and metre.

If W is the total load on the beam,

$$W = wL$$

Because $W = R_A + R_B$, from symmetry,

$$R_A = R_B = \frac{W}{2}$$

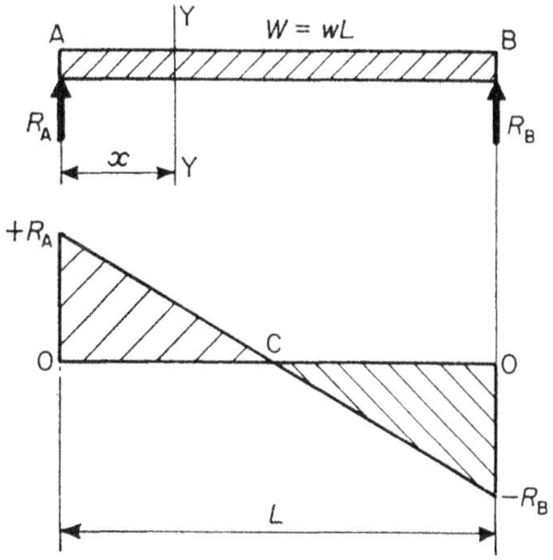

Figure 10.10 Shearing force diagram—simply supported beam with uniformly distributed load.

At A the shearing force is $+R_A$, and so, having chosen a suitable scale, plot $+R_A$ vertically above the datum line at O.

Between A and B a downward, and hence a negative, force is applied continuously, at the rate of w for every unit length of travel to the right from A. Therefore, when C, the mid-span position, is reached half the total load, $\left(\text{or } \frac{W}{2} = \frac{wL}{2}\right)$ has been applied downwards.

Therefore, at C the shearing force has a value of $\left(+R_A - \frac{W}{2}\right)$, which is zero.

As the value of x from A increases, the downward load on the portion of the beam to the left of the section YY increases proportionately, and therefore the downward inclination of the shearing force diagram between A and B is constant.

Consequently, the value of the shearing force when B is reached is $(+R_A - W)$ which, is equal to $-R_B$.

At B a positive shearing force of R_B is added for the reaction at B, bringing the diagram back to the zero line.

Note that the value of the shearing force at any section along the beam is determined by the height of the shearing force diagram at that section.

2. *Cantilever*

Concentrated load

Consider a cantilever with a single load concentrated at the free end of the beam. Take any section YY at a distance x from the left hand extremity of the beam; the shearing force at that section is $-W$, because the force to the left of the section is acting downwards.

No other external force acts on the beam between the free end and the fixed end.

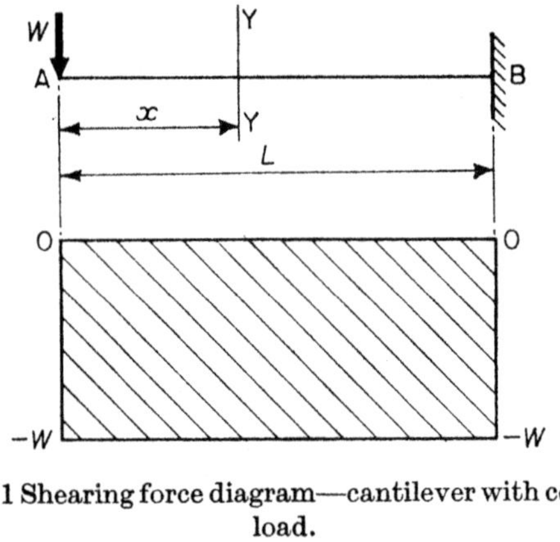

Figure 10.11 Shearing force diagram—cantilever with concentrated load.

To draw the shearing force diagram, select a suitable scale, and plot W below the zero line to represent the shearing force at A. Between A and B the shearing force remains constant at—W. At B, the support reaction, acting upwards, brings the diagram back to the zero line.

Distributed load

Consider a cantilever with a load uniformly distributed over its whole length; thus $W = wL$ where w = load per unit length of beam. At a distance x from the free end of the beam the total downward load is wx, and therefore the shearing force is $-wx$.

Because the shearing force is proportional to the distance from A the graph of the shearing force is a straight line inclined downwards from a value of O at A (where $x = O$) to a value of $-wL$ or $-W$ at B (where x is equal to L).

At B an upward reaction W, brings the diagram back to the zero line.

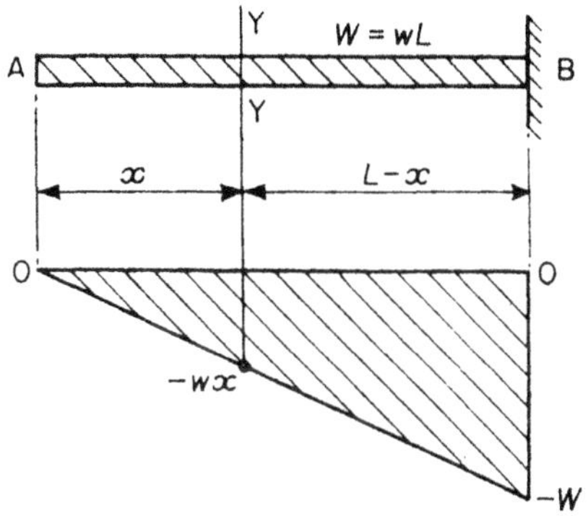

Figure 10.12 Shearing force diagram—cantilever with uniformly distributed load.

Summary

1. In a shearing force diagram all lines are straight if the loading is concentrated or uniformly distributed.
2. For concentrated loads the diagram consists of horizontal straight lines joined by vertical lines.
3. For a uniformly distributed load the diagram consists of one inclined straight line terminating in a vertical line.
4. The vertical height of the diagram at a distance x from the left hand extremity of the beam represents to scale the shearing force at that section. Above the datum line the shearing force is positive, below it is negative.

(*Note:* In practice it often happens that a beam carries both concentrated and distributed loads. In this case the shearing force diagram is a combination of the diagram types of points 2 and 3 in the summary.)

Bending moment

Consider again a simply supported beam with a single central load, as in figure 10.13.

Take a vertical section YY at a distance x from the left hand extremity of the beam. There is a tendency for the left hand portion of the beam to rotate in a clockwise direction about YY because of the force R_A.

Now the moment of a force about a point is equal to the product of the force and the perpendicular distance between the line of action of the force and the point under consideration.

The moment of the force R_A about the YY line is equal to $R_A x$. This moment, because it is tending to bend the beam, is called *the bending moment about* YY; stresses are induced in the beam opposing this tendency.

For equilibrium of the beam the clockwise bending moment on the

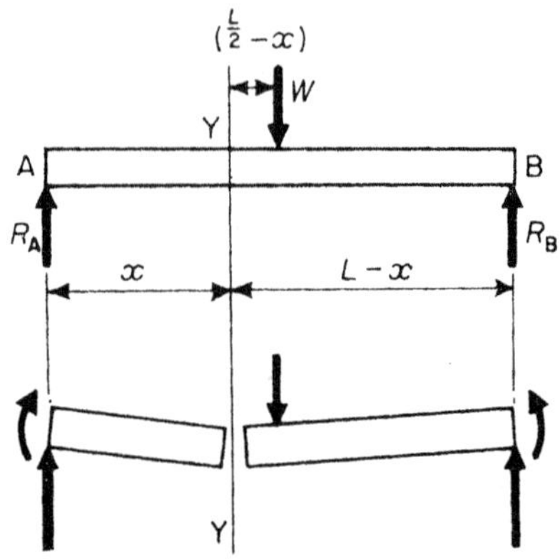

Figure 10.13 The bending of a beam under load.

left hand side of the section YY must be balanced by an anti-clockwise bending moment on the right hand side of the section; this must be the net result of an anti-clockwise moment equal to $R_B(L - x)$ and a clockwise moment of $W\left(\frac{L}{2} - x\right)$ i.e. $R_B(L - x) - W\left(\frac{L}{2} - x\right)$.

Under more complex loading there will be several forces tending to bend the beam in opposite directions on either side of a chosen section.

Therefore, in general, the bending moment at any section of a beam is equal to the algebraic sum (that is, taking clockwise and anti-clockwise directions into account) of the moments of all the external forces (including the reactions) acting on the beam *either* to the left *or* to the right of the section chosen, about the section.

Convention

Measuring x from the left to right, as before, a clockwise moment on the left of the section is considered to be positive.

Such a moment, must be balanced by an anti-clockwise moment to the right of the section. These two moments cause the beam to sag, and so they are called 'sagging moments'. The effect is shown diagrammatically, and greatly exaggerated, in figure 10.14.

If, however, the directions of the external forces were reversed, the

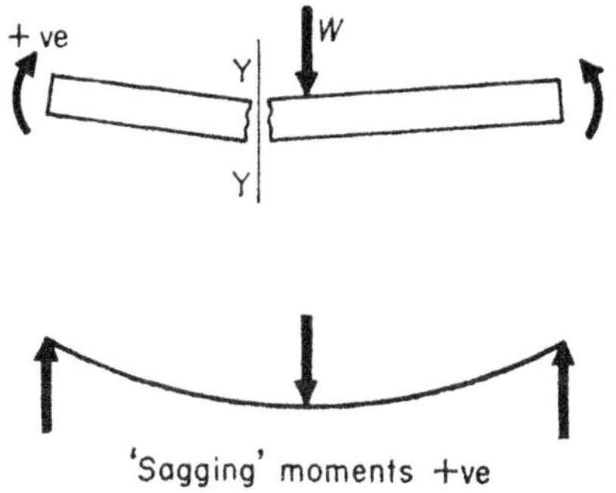

Figure 10.14 Sagging moments.

moment on the left hand portion of the beam in figure 10.15 would be anti-clockwise, and thus considered to be a negative bending moment. This moment would be balanced by a clockwise moment of equal magnitude acting on the right hand portion of the beam, causing it to have the appearance of a humped back. These moments are called

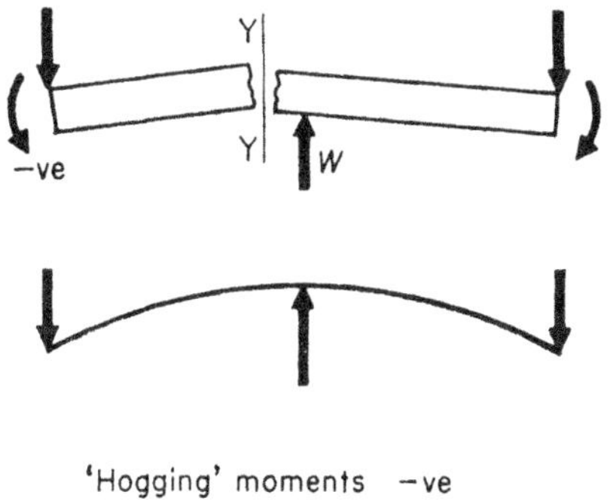

Figure 10.15 Hogging moments.

'hogging moments'. The effect is shown diagrammatically in figure 10.15.

Convention. Sagging moments may be taken as *positive.* In which case hogging moments will be negative.

Bending moment diagrams

A bending moment diagram is a graph showing the variation of the bending moment along the length of the beam. As with a shearing force diagram it is an advantage to project it below the space diagram. If, however, both shearing force and bending moment diagrams are required, then the bending moment diagram is usually drawn below the shearing force diagram, which is drawn immediately below the space diagram.

It should be drawn to scale, and the scale stated in the correct units.

Simply supported beam

(a) Concentrated load

Consider a simply supported beam with a single concentrated load at its mid-span.

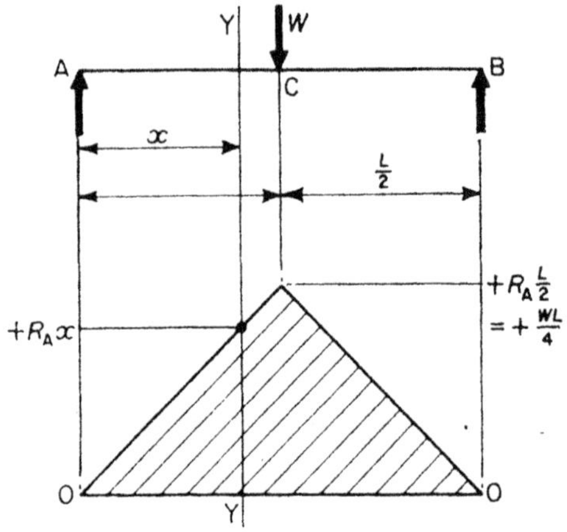

Figure 10.16 Bending moment diagram—simply supported beam with concentrated load.

At section YY a distance x from the left hand extremity of the beam the bending moment of the only force—R_A—to the left of the section YY is $R_A x$ and, because it is clockwise, the moment is positive and a sagging moment.

The magnitude of the moment increases from O to $R_A \frac{L}{2}$ as the section YY moves from A (where $x = O$) to C $\left(\text{where } x = \frac{L}{2}\right)$. From symmetry,

$$R_A = R_B = \frac{W}{2}$$

Therefore $$\text{Bending moment at C} = \frac{WL}{4}$$

The value of the bending moment depending entirely on a linear increase of x must be represented by a continuous straight line between A and C, as shown in figure 10.16, inclined upwards to represent a positive bending moment. By reason of symmetry, bending moments on the right hand portion of the beam must be equal to the moments on the left hand portion for equal distance from the extremities of the beam.

The diagram can therefore be completed without further calculation by drawing a line inclined downwards to meet the zero line at the right hand extremity.

Note that the units of the bending moment are (because moment equals force × distance) lbf in., lbf ft, tonf in. or tonf ft or, in metric units, kgf m.

Note also that the height of the diagram at any section of the beam represents to scale the bending moment at that section, and that at the extremities of the beam the bending moment is zero.

(*b*) *Distributed load*

Now consider a simply supported beam with a uniformly distributed load over its whole span.

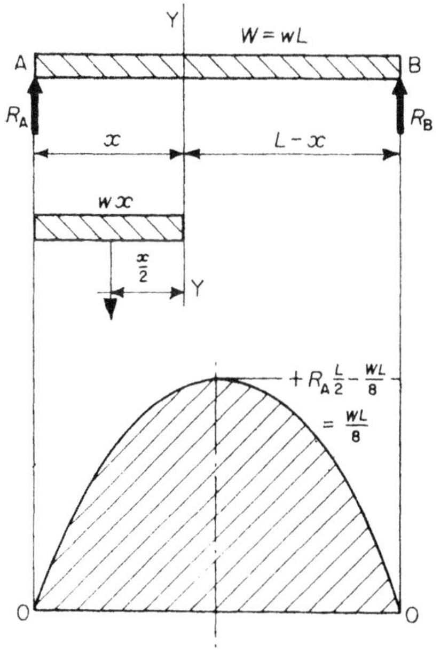

Figure 10.17 Bending moment diagram—simply supported beam with uniformly distributed load.

As before, consider a section YY at a distance x from the left hand extremity of the beam. In figure 10.17 the distributed load over the portion of the beam to the left of YY is wx and the distance of the mid-point of the load from YY is $\frac{x}{2}$.

For the purpose of taking moments, regard a distributed load over the left hand portion of the beam—that is, a load of wx—as equivalent to a single concentrated load of wx acting at the centre of the section of length x. To the left of YY,

Clockwise (and therefore positive) moment about $YY = R_A x$

Anti-clockwise (and therefore negative) moment about YY

$$= wx \cdot \left(\frac{x}{2}\right) = \frac{wx^2}{2}$$

The clockwise moment is proportional to x, and could therefore be represented separately by a straight line. The anti-clockwise moment is proportional to x^2 and would be represented separately by a curved line.

The combined, or net, effect of the two moments is also represented by a curved line *above* the zero line, because the positive, or sagging moment, is greater than the negative, or hogging moment.

A curved line is not as easy to draw as a straight line, and so it will be necessary to evaluate the bending moment at several sections—usually equally spaced—along the span.

Clearly, at the mid-span position, where $x = \frac{L}{2}$, the bending moment is

$$+R_A \frac{L}{2} - \frac{w}{2}\left(\frac{L}{2}\right)^2 = +\frac{R_A L}{2} - \frac{WL}{8} \quad \text{because} \quad W = wL$$

From symmetry $R_A = R_B = \frac{W}{2}$, and therefore the bending moment at mid-span is $\frac{WL}{8}$.

Cantilever

(a) Concentrated load

A cantilever is loaded at its free end with a single concentrated load, as in figure 10.18.

The bending moment of the downward force W about the section YY is $-Wx$, an anti-clockwise or hogging moment.

At A, when x is equal to zero, the bending moment is zero. Because the bending moment is proportional to the distance x from the free end, the bending moment at B will be $-WL$. The graph will be a straight line inclined downwards from a value O at A to a value $-WL$ at B.

Note that, whereas the bending moment is zero at the free end of the beam, it has a finite value at an end that is fixed.

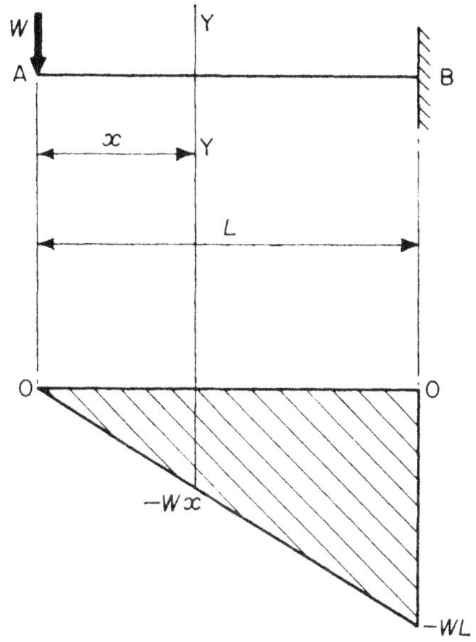

Figure 10.18 Bending moment diagram—cantilever with concentrated load at free end.

(b) Distributed load

A cantilever with a load uniformly distributed over its whole length, is illustrated in figure 10.19.

$$\text{Total load, } W = wL$$

w being the load per unit length of beam. At section YY a distance of x from the free end of the beam the load on the portion of the beam to the left of YY is wx, and its moment about YY is $wx\left(\frac{x}{2}\right)$.

This moment is a hogging moment and therefore negative.

At A, where x is zero, the bending moment is zero. Between A and B the moment varies in proportion to x^2 and therefore the graph is a curved line bending downwards at an increasing rate as the distance x is increased.

At B, the fixed end, the value of the bending moment is

$$-\frac{wL^2}{2} = -\frac{WL}{2}$$

(*Note:* There is no merit in trying to remember the maximum values of the bending moment for each of the four cases just considered. It is much more important to learn the basic principles.)

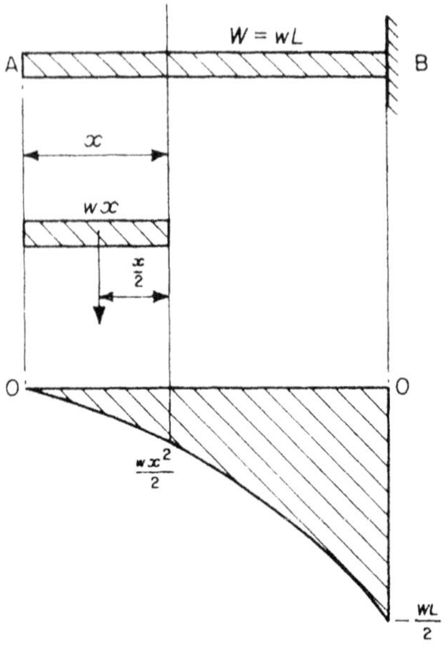

Figure 10.19 Bending moment diagram—cantilever with uniformly distributed load.

Summary

1. A bending moment diagram may consist of either straight lines or curved lines, or a combination of the two.

2. For concentrated loads the diagram consists of inclined straight lines, the bending moment at extremities which are 'free' having always a zero value.

3. For a distributed load the diagram consists of a curved line. Again the bending moment at 'free' extremities is zero.

4. The vertical height of the diagram at a section represents to scale the bending moment at that section. Above the datum (or zero) line the bending moment is positive; below it is negative.

Calculation of the magnitudes of the reactions

This has already been dealt with incidentally (p. 129) but is repeated here because of its importance.

For the purpose of explaining the meaning of shearing force and bending moment, examples of simply supported beams have been chosen in which the loads were symmetrical on either side of a central vertical axis. In these examples it can be argued, merely from visual observation, or from symmetry, that the values of the reactions at the supports are equal. It is not often that such simple cases arise, and so it becomes necessary frequently to calculate the values of the reactions. This must always be done when the load distribution is not uniform.

In order to calculate these values consider the equilibrium of the whole beam.

Because all the external forces on the beam are vertical, and therefore parallel, the conditions of equilibrium are:

1. All the vertical forces must balance. That is, the sum of all the upward forces must equal the sum of all the downward forces.
2. The algebraic sum of the moments of all the forces about any point must be zero.

Here it is necessary to emphasize the distinction between the calculation of a bending moment at any point and taking moments about the point.

The calculation of a *bending moment* involves adding up all the clockwise moments and subtracting from this total the sum of all the anti-clockwise moments of forces which are acting on ONE SIDE ONLY of the point or section chosen. This calculation gives a net, or effective, moment of all the forces on one side of the section.

To *take moments* about a point, the equilibrium of *both sides* of the section chosen is being considered, in which case, because the net, or effective, moments on one side must balance the net, or effective, moments on the other, it is clear that

> 'The algebraic sum of the moments of all the forces on both sides of the section, about the section, is zero.'

In fact, this is stating an equation.

The moment of a force about a point which is *on* the line of action of the force is zero. Consequently, by selecting a point on the line of action of one of the two support reactions, that reaction is eliminated from our moment equation.

Therefore, in order to calculate the value of a support reaction, take moments of all the external forces, including the reactions, about the second support and equate the totals of the clockwise and anti-clockwise moments.

The value of *each* reaction should be calculated in turn, and then the first condition of equilibrium, namely

'All the vertical forces, including the reactions, must balance' can be used to check that the calculations have been made correctly.

Thus $$R_A + R_B = \Sigma(W)$$

where Σ (Greek letter sigma),

= Algebraic sum of all external loads on the beam

Example

A beam 20 ft long is simply supported at its ends. It is loaded with concentrated loads of 2 tonf, 5 tonf and 3 tonf at distances of 4 ft, 8 ft and 15 ft respectively from the left hand extremity. Draw to scale the shearing force and bending moment diagrams.

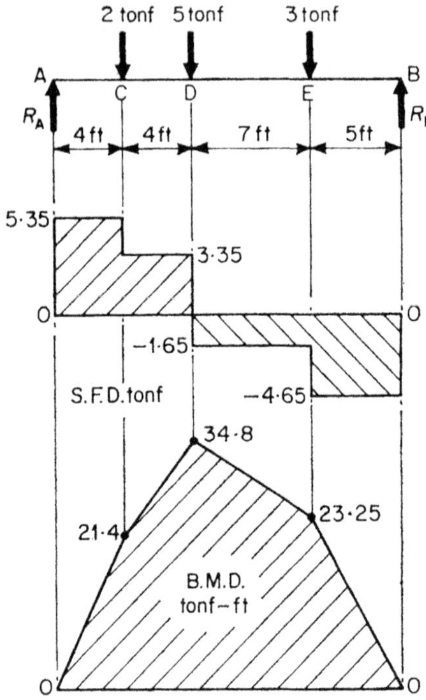

Figure 10.20

(SI. If dimensions were in kilonewtons and metres, above figures would apply to these magnitudes.)

Solution

First of all determine the magnitude of the reactions by taking moments about each end of the beam in turn.

Moments about A

$$\text{Clockwise moments} = \text{Anti-clockwise moments}$$

$$(2 \times 4) + (5 \times 8) + (3 \times 15) = (R_B \times 20)$$

$$8 + 40 + 45 = 20R_B$$

$$93 = 20R_B$$

$$\underline{R_B = 4{\cdot}65 \text{ tonf}}$$

Moments about B

$$(R_A \times 20) = (3 \times 5) + (5 \times 12) + (2 \times 16)$$

$$20R_A = 15 + 60 + 32$$

$$= 107$$

$$\underline{R_A = 5{\cdot}35 \text{ tonf}}$$

Check

$$R_A + R_B = 10 \text{ tonf} = \text{Sum of loads}$$

Shearing forces

Between A and C S.F. $= +R_A = +5{\cdot}35$ tonf

Between C and D S.F. $= +5{\cdot}35 - 2 = +3{\cdot}35$ tonf

Between D and E S.F. $= +3{\cdot}35 - 5 = -1{\cdot}65$ tonf

Between E and B S.F. $= -1{\cdot}65 - 3 = -4{\cdot}65$ tonf

Bending moments

At A Bending moment $= 0$

At C B.M. $= +(R_A \times 4) = +(5{\cdot}35 \times 4) = +21{\cdot}4$ tonf-ft

At D B.M. $= +(5{\cdot}35 \times 8) - (2 \times 4) = +42{\cdot}8 - 8 = +34{\cdot}8$ tonf ft

At E B.M. $= +(5{\cdot}35 \times 15) - (2 \times 11) - (5 \times 7)$

$= +80{\cdot}25 - 22 - 35$

$= +80{\cdot}25 - 57 = +23{\cdot}25$ tonf ft

At B B.M. $= 0$

Note that values have been placed against each change point. Set out your diagrams and your calculations properly and neatly.

Watch the units closely and always state them with the numerical answers.

Subsequent worked examples will be covered in less detail but essential steps are given; the student is recommended to work out the details independently.

Example

A cantilever beam carries three point loads as shown in the figure. Calculate the values of shearing force in the beam between the load positions and the values of bending moment at the load positions. Draw scale diagrams to show the variation of shearing force and bending moment at all sections of the beam.

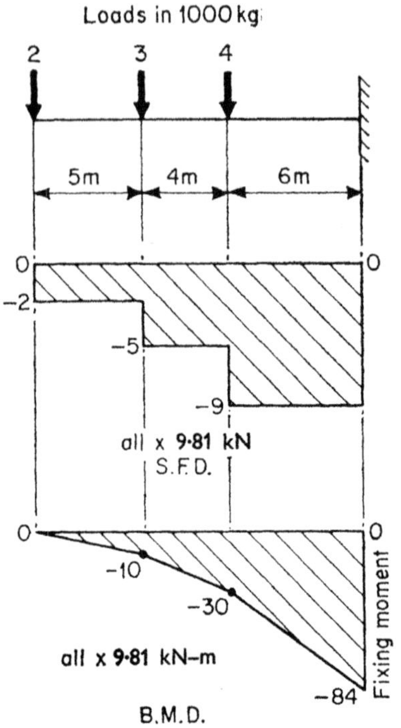

Figure 10.21

Solution

Using the left hand, free, end of the beam as the zero for section positions and S and M as symbol letters for shearing force and bending moment respectively.

$$S_0 \text{ to } S_5 = -2 \times 9{\cdot}81 \text{ kN}$$
$$S_5 \text{ to } S_9 = -2 - 3 = -5 \text{ (all} \times 9{\cdot}81 \text{ kN)}$$
$$S_9 \text{ to } S_{15} = -2 - 3 - 4 = -9 \text{ (all} \times 9{\cdot}81 \text{ kN)}$$
$$M_0 = 0 \text{ —m}$$
$$M_5 = -2 \times 5 = -10 \text{ (all} \times 9{\cdot}81 \text{ kN —m)}$$
$$M_9 = -2 \times 9 - 3 \times 4 = -30 \text{ (all} \times 9{\cdot}81 \text{ kN —m)}$$
$$M_{15} = -2 \times 15 - 3 \times 10 - 4 \times 6 = -84 \text{ (all} \times 9{\cdot}81 \text{ kN —m)}$$

Example

A beam, 16 ft long, is simply supported at its ends and carries a load of 1 ton per ft over its whole length. Calculate the values of shearing force and bending moment at a section 1 ft from the left hand end and at 2 ft intervals along the beam.

Plot these values and sketch the complete shearing force and bending moment diagrams for the beam.

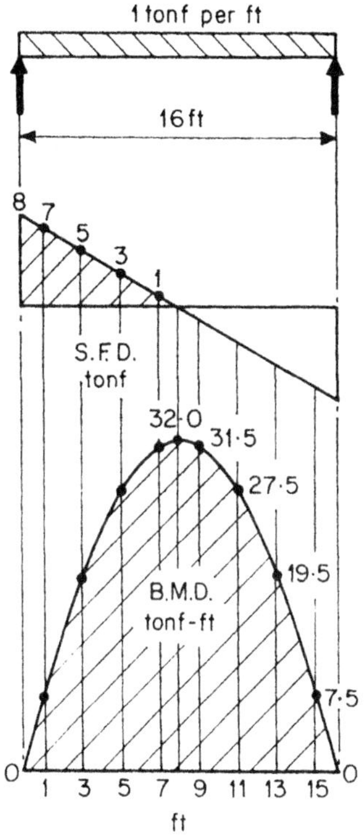

Figure 10.22

Solution

The two support reaction are equal and each has a value of 8 ton.

$$S_1 = +8 - 1 = +7 \text{ tonf} \qquad M_1 = 8 \times 1 - 1 \times 0{\cdot}5 = +\ 7{\cdot}5 \text{ tonf ft}$$

$$S_3 = +8 - 3 = +5 \text{ tonf} \qquad M_3 = 8 \times 3 - 3 \times 1{\cdot}5 = +19{\cdot}5 \text{ tonf ft}$$

$$S_5 = +8 - 5 = +3 \text{ tonf} \qquad M_5 = 8 \times 5 - 5 \times 2{\cdot}5 = +27{\cdot}5 \text{ tonf ft}$$

$$S_7 = +8 - 7 = +1 \text{ tonf} \qquad M_7 = 8 \times 7 - 7 \times 3{\cdot}5 = +31{\cdot}5 \text{ tonf ft}$$

$$S_9 = +8 - 9 = -1 \text{ tonf} \qquad M_9 = 8 \times 9 - 9 \times 4{\cdot}5 - +31{\cdot}5 \text{ tonf ft}$$

$$S_{11} = +8 - 11 = -3 \text{ tonf} \qquad M_{11} = 8 \times 11 - 11 \times 5{\cdot}5 = +27{\cdot}5 \text{ tonf ft}$$

$$S_{13} = +8 - 13 = -5 \text{ tonf} \qquad M_{13} = 8 \times 13 - 13 \times 6{\cdot}5 = +19{\cdot}5 \text{ tonf ft}$$

$$S_{15} = +8 - 15 = -7 \text{ tonf} \qquad M_{15} = 8 \times 15 - 15 \times 7{\cdot}5 = +\ 7{\cdot}5 \text{ tonf ft}$$

(*Note:* When loading is uniformly distributed the bending moment varies in a non-linear way; the variation is represented by a curve. The shearing force however is still represented by a straight line but now this is inclined to the base line.)

Two principles are worth noting; they are an aid to drawing the diagrams, particularly in complex examples.

First, *peak values of bending moment occur at sections where the shearing force is zero or changes from a positive to a negative value.*

Secondly, *the difference in bending moment at any two sections is equal to the area of the shearing force diagram between those sections.*

Apply these principles to the previous example. Shearing force is zero at the centre of the beam; the peak, or maximum, value of bending moment must occur at the mid-section. At the left-hand support the shearing force is +8 ton and the bending moment zero.

Between the support and mid-section, the area of the shearing force diagram—a triangle of height + 8 tonf and base length 8 ft—is $\frac{1}{2} \times (+8 \text{ tonf}) \times 8$ ft or 32 tonf-ft.

This must be the bending moment value at mid-section since that at the end is zero.

The principles may be verified from an examination of the following two worked examples.

Example

A beam ABCDE is simply supported at A and E.

$$AB = 7{\cdot}5 \text{ ft},\quad BC = 5 \text{ ft},\quad CD = 7{\cdot}5 \text{ ft} \quad \text{and} \quad DE = 5 \text{ ft}$$

It carries downward loads of 10 and 4 tons respectively at B and D and a vertical upward load of 6 tons at C. Find the reactions at A and E. Find the bending moments at B, C and D. Draw to scale the bending moment and shearing force diagrams for the beam using the following scales: length 1 in. to 5 ft; bending moment, 1 in. to 10 tonf-ft; shearing force, 1 in. to 4 tons. (U.L.C.I.) May 1961

Solution

Taking moments about A.

$$R_E \times 25 = 10 \times 7{\cdot}5 - 6 \times 12{\cdot}5 + 4 \times 20$$
$$R_E = 3{\cdot}2 \text{ tonf}$$

Taking moments about E.

$$R_A \times 25 = 10 \times 17{\cdot}5 - 6 \times 12{\cdot}5 + 4 \times 5$$
$$R_A = 4{\cdot}8 \text{ tonf}$$

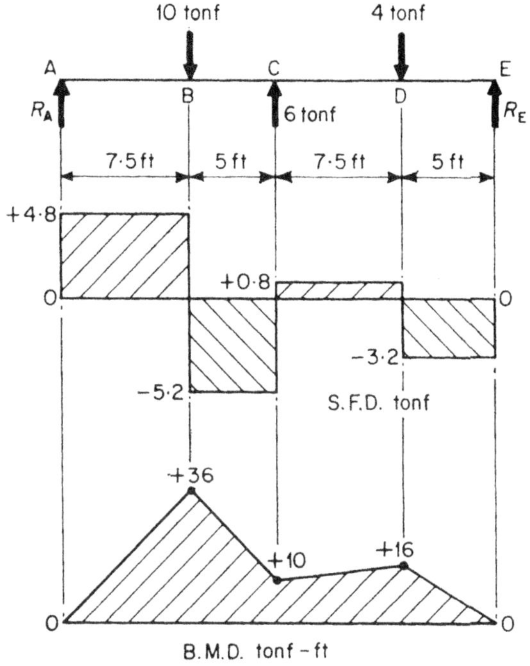

Figure 10.23

Comments. The peak values of bending moment occur at sections B and D where the shearing force changes sign.

The areas of the shearing force diagram between sections A–B, B–C, C–D and D–E are +36, −26, +6 and −16 tonf ft respectively. By starting at A and adding in succession these areas, the values of bending moment at B, C, D and E will be determined.

Example

A cantilever beam ABCDE is fixed horizontally at A.

AB = 5 ft, BC = 9 ft, CD = 6 ft and DE = 5 ft

It carries vertical downward loads of 10, 5 and 4 tons respectively at B, C and E. If a vertical upward load W tons acts at D, find the value of W which reduces the numerical value of the bending moment at A to 60 ton ft. Find the bending moments at D, C and B. Draw to scale the resulting bending moment and shearing force diagrams for the beam using the following scales: length, 1 in. to 5 ft; bending moment, 1 in. to 20 ton ft; shearing force, 1 in. to 5 tons.

(U.L.C.I.) June 1961

Solution

This is an example with a difference.

First calculate the value of the bending moment at A if the upward load at D is removed; this will be equal to $10 \times 5 + 5 \times 14 + 4 \times 25$ or 220 tonf ft clockwise.

The bending moment due to the upward load W at D, acting alone, is $W \times 20$ tonf ft anticlockwise. Then

$$220 - W \times 20 = 60 \text{ tonf ft}$$

Therefore
$$W \times 20 = 160 \text{ tonf ft}$$
$$\underline{W = 8 \text{ tonf}}$$

The vertical reaction at A must be 11 tonf upwards.

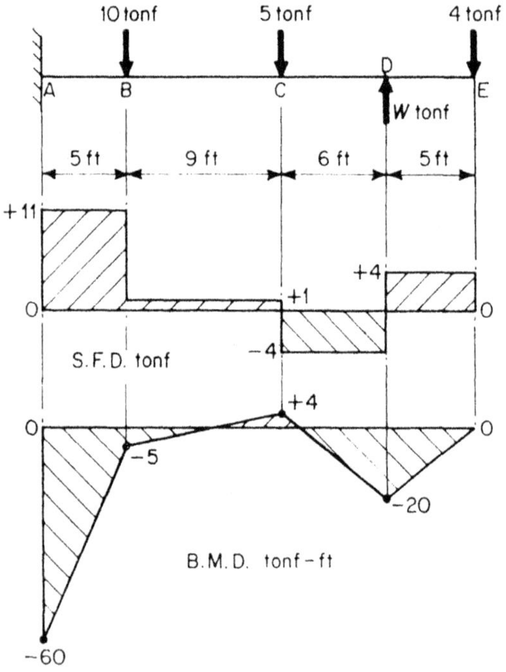

Figure 10.24

Comments. It will be noted that at sections C and D, where the shearing force changes sign, there are peak values of bending moment, neither an absolute maximum.

Areas of the shearing force diagram A–B, B–C, C–D and D–E are +55, +9, −24 and +20 tonf ft, respectively. Starting at A with a bending moment of −60 tonf ft, the values at B, C, D and E are −5, +4, −20 and 0 tonf ft, in that order.

Summary

Beams. Types of beam, and loading

Beams can be simply supported, in which case the ends are unconstrained, or have their ends built into the supports. The latter type includes the cantilever beam, where one end is built in.

Loading can be *concentrated* or *distributed.* At this stage of the course, only uniformly distributed loads have been considered.

Shearing force

This is *the sum of all the forces acting across the section of the beam, on one side of the point under consideration.*

A shearing force diagram is a graph showing the variation of shearing force along the beam. Some convention is used for drawing such a diagram, to denote positive and negative values of shearing force. Where the left hand section is caused to tend to slide upwards relative to the right hand section, the shearing force is considered to be *positive.*

Bending moment

This is *the sum of the moments of all the forces on one side of the section of the beam under consideration.*

A bending moment diagram is a graph showing the variation of bending moment along the beam. Moments which cause 'hogging' of the beam have been taken as negative, and 'sagging' moments have been taken as positive.

Two principles are worth noting here:

(*a*) Peak values of bending moment occur at sections where the shearing force is zero, or changes from a positive to a negative value or vice versa.

(*b*) The difference between the bending moments at any two sections is equal to the area of the shearing force diagram between the sections.

Exercises

1. Draw, to scale, the shearing force and bending moment diagrams for a simply supported beam of 20 ft span carrying the following concentrated loads: 4, 5, 9, 3 and 10 cwt at 3, 7, 10, 14 and 18 ft respectively from the left hand support.

2. A beam, 20 metres long, is simply supported at its ends. It carries concentrated loads of 3, 5 and 3 metric tons at distances of 5, 8 and 15 metres respectively from the left hand support. Draw, to scale, the S.F. and B.M. diagrams. Determine also the maximum bending moment and state the position where it occurs.

3. A beam, 25 ft long, is supported at its ends. Vertical loads of 3, 5 and 4 tonf are carried at points 5, 10 and 15 ft respectively from the left hand support. Draw, to scale, the S.F. and B.M. diagrams. State the position and magnitude of the maximum bending moment.

4. A steel girder of 30 metres span is simply supported at its ends, and carries loads of 5, 3 and 5 by 1000 kg at, 8, 12 and 25 metres respectively from the left hand support. Draw, to scale, the S.F. and B.M. diagrams. From these diagrams determine the values of the S.F. and B.M. at mid span.

5. A weightless beam, 10 ft long, carries a central load of 6 tonf and a load of 2 tonf at each end. It is simply supported on supports which are 6 ft apart. Draw, to scale, the B.M. diagrams for the beam, giving all the principal values:

(*a*) when the loaded beam is placed symmetrically on the supports.
(*b*) when the loaded beam is moved 1 ft from the position of symmetry.

6. A beam, 8 metres long, is supported at positions 1 metre and 6·5 metres from the left hand end. It carries concentrated loads of 250, 300 and 250 kg at 2, 4 and 8 metres respectively from the left hand end. Draw, to scale, the S.F. and B.M. diagrams. From these diagrams determine the S.F. and B.M. at a position 5 metres from the left hand end, and check the values by calculation.

7. A uniform beam, 20 ft long, is supported horizontally by two supports at distances 3 ft and 18 ft from the left hand end. It carries concentrated loads of 3, 5 and 4 tonf at distances of 5, 15 and 20 ft respectively from the left hand end. Determine the reactions at the supports, and draw to scale the S.F. and B.M. diagrams. Hence obtain the magnitude and position of the maximum B.M.

8. A beam, 10 metres long, is simply supported at its ends, and is uniformly loaded at 60 kg per metre run. Draw the S.F. and B.M. diagrams and calculate the values of the S.F. and B.M. at positions 2 and 4 metres from the left hand support.

9. A cantilever projects 10 ft from a wall and carries a load of 8 cwt at a distance of 2 ft from the wall, and another load of 5 cwt at its free end. Draw, to scale, the S.F. and B.M. diagrams and from these determine the shearing force and bending moment at (*a*) the wall (*b*) a point 4 ft from the wall.

10. A cantilever is 3 metres long and is loaded uniformly with 4,000 kg per metre run. Draw the S.F. and B.M. diagrams.

Calculate the value of the B.M. at a point 0·75 metre from the free end.

11

Centres of gravity and centroids

It was seen in Chapter 3 that the Earth exerts a measurable force of attraction on all bodies towards its centre. Consider a solid body to be made of a very large number of small bodies or particles; the Earth attracts each one of these particles towards its centre by what is known as the force of gravity. Each gravitational force is known as the weight of the particle. Because the Earth is so large in comparison with any body on it the radial lines of force attracting each particle to the Earths centre can be regarded as parallel to each other. If all these tiny gravitational forces are added together the total represents the weight of the body itself. In other words, the resultant of all the gravitational forces acting on the tiny particles is equal to the weight of the body.

Centre of gravity

Suppose that a solid body of simple rectangular section lies on one face as shown in figure 11.1. Imagine the body to be divided up into a very large number of vertical sections, and that the pull of the Earth's gravitational field can be represented by the small arrows acting vertically downwards from the whole of the base of the body.

The resultant force is one which produces the same effect as all these tiny gravitational forces acting together, and its line of action can be represented by the thick vertical line.

Now consider the body in a new position such that it lies flat on another face, the value of the resultant force remains the same, and equal to the weight of the body.

Suppose the new line of action of the resultant force is represented

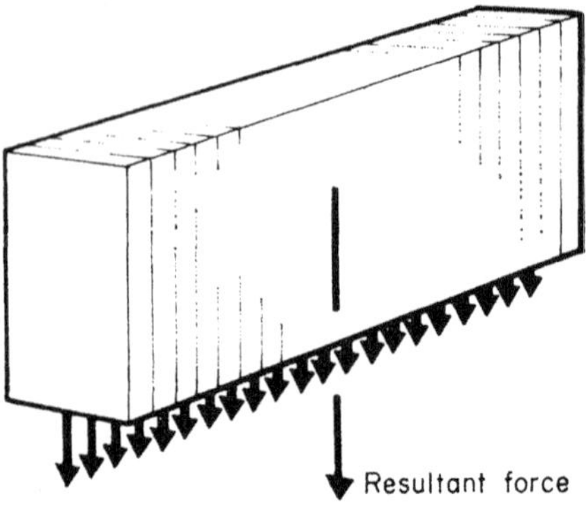

Figure 11.1

by the thick vertical line in figure 11.2 and the original line of action by a thick horizontal line. These two lines can be shown to intersect at a point G. Thus, for any position of the body—whether it is lying on its back, on its side, or on its end—it can be shown experimentally that the resultant force acts through the intersection of all the respective lines of action. The intersection must be a single point, and is known as the *centre of gravity* of the body.

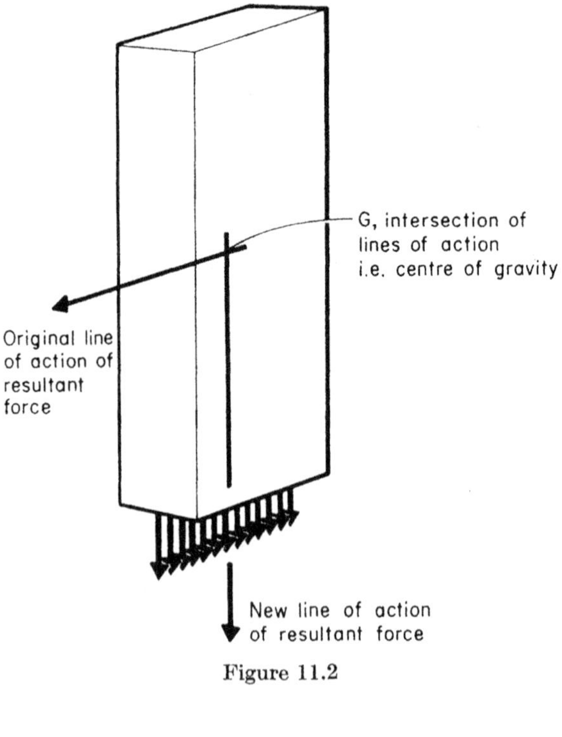

Figure 11.2

It is assumed that the weight of a body acts vertically downwards through its centre of gravity. The centres of gravity of many bodies can be determined merely by inspection because they will be centres of symmetry. Thus, the centre of gravity of a sphere is at its centre, and the centre of gravity of a solid cube is at the intersection of the long, or cross, diagonals.

If the resultant force—which is the weight—has the same effect as all the gravitational forces acting on the small particles which make up the whole body, it can be stated that the moment of the resultant about any point must be equal to the sum of the moments of all the gravitational forces about the same point.

This statement may be expressed in terms of an equation, thus

$$W\bar{x} = w_1x_1 + w_2x_2 + w_3x_3 + w_4x_4 + \ldots w_nx_n$$

where $x_1, x_2, x_3, \ldots x_n$ represent the distances of the gravitational forces $w_1, w_2, w_3, \ldots w_n$ from the fixed point, and

W = Resultant of all the gravitational forces acting on the small particles,
$= w_1 + w_2 + w_3 + \ldots w_n$,
$\bar{x}$ = Length of perpendicular from the fixed point to the line of action of the resultant.

Note that $\bar{x}$ and $\bar{y}$ are the usual forms of writing down the distances from the vertical Y axis and the horizontal X axis respectively of a *resultant* force.

The above equation may be used to determine the position of the centre of gravity of awkward shaped bodies built up from symmetrical shapes whose centres of gravity coincide with their centres of symmetry.

Take the simple case of a cylindrical shaft which has been turned down at one end to a smaller diameter, and determine its centre of gravity.

Referring to figure 11.3 the larger end A of the shaft has a diameter d_1 and length l_1, and the smaller end B has a diameter d_2 and length l_2. Let the weight of A be w_1 and that of portion B be w_2.

By inspection, the centre of gravity of the complete shaft must lie on the axis of symmetry XX, as do the centres of gravity of the symmetrical cylinders A and B. Further the weights of A and B must act through points which are mid-way along their respective lengths. That is,

$$x_1 = \tfrac{1}{2}l_1 \quad \text{and} \quad x_2 = l_1 + \frac{l_2}{2}$$

where x_1 and x_2 are measured from the left hand end of the shaft denoted by the vertical line YY.

Let G = Centre of gravity of the complete shaft of weight W, and

$\bar{x}$ = Distance along the axis of symmetry XX of G from vertical line YY.

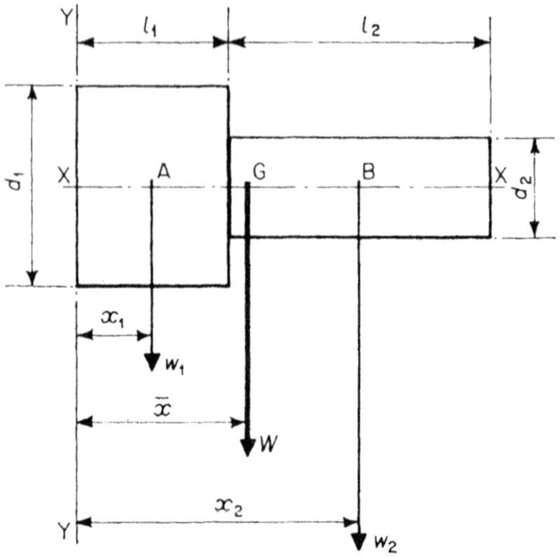

Figure 11.3

Take moments about the line YY

$$W\bar{x} = w_1x_1 + w_2x_2$$

i.e.
$$\bar{x} = \frac{w_1x_1 + w_2x_2}{W}$$

or,
$$\frac{w_1x_1 + w_2x_2}{w_1 + w_2}$$

which, by substitution, allows the determination of $\bar{x}$.

The last expression may be written in a more general form thus:

$$\bar{x} = \frac{\text{Sum of all the moments of weights about } YY}{\text{Sum of all the weights}}$$

$$= \frac{\Sigma wx}{\Sigma w}$$

where Σ (Greek letter sigma) is the mathematical sign denoting "sum of".

Centroid

If the body takes the form of a flat thin sheet the centre of gravity can be determined quite readily in the laboratory by suspending it from several different points near its edge. If vertical lines are drawn through the points of suspension in turn by using a plumb bob as a guide it will be found that all the lines intersect at a single point.

Assume the thickness of the sheet, or lamina as it is sometimes called, to be constant and the material to have the same specific weight throughout. The point of intersection then corresponds to a centre of area, known as the *centroid* of the area. Now

$$\text{Specific weight} = \frac{\text{Weight}}{\text{Volume}}$$

$$= \frac{\text{Weight}}{\text{Surface area of lamina} \times \text{Thickness}}$$

Therefore, since the specific weight and the thickness of the lamina are uniform,

$$\text{Weight of lamina} = \text{Constant} \times \text{Surface area}$$

For laminae or thin surfaces, the equation for $\bar{x}$ may be written

$$\bar{x} = \frac{(a_1x_1 + a_2x_2) \times \text{Constant}}{A \times \text{Constant}}$$

where $\text{Constant} = \text{Specific weight} \times \text{Thickness}$

$$= \frac{a_1x_1 + a_2x_2}{A}$$

or, in more general terms

$$A\bar{x} = a_1x_1 + a_2x_2 + a_3x_3 + \ldots a_nx_n$$

where $A = a_1 + a_2 + a_3 + \ldots a_n$

and $\bar{x}$ = Distance of the centroid of the surface A from a fixed line

$$\bar{x} = \frac{\Sigma ax}{\Sigma a}$$

As it is possible to determine the centres of gravity of some solids by inspection, so it is possible to determine the centroids of some areas. Thus the centroid of a circle is at its geometrical centre, and the centroid of a rectangle is at the intersection of its diagonals. The centroid of a triangle is at the intersection of its medians; a median being a line drawn from a corner (or apex) of a triangle to the mid-point of the opposite side. This point of intersection can also be defined in terms of the vertical height of the triangle if one side—the base—is horizontal, since its distance from the base is one third of the vertical height.

Tabular solution of problems

The above general equations are of such a form that the calculations of values for $\bar{x}$ and $\bar{y}$ can be made easily and conveniently in a tabular form either for a composite solid or for an area built up from a number of standard shapes, as illustrated below.

Determination of position of Centre of Gravity

Part	*Weight*	*Distance of C of G from YY*	*Moment about YY*	*Distance of C of G from XX*	*Moment about XX*
1	w_1	x_1	w_1x_1	y_1	w_1y_1
2	w_2	x_2	w_2x_2	y_2	w_2y_2
3	w_3	x_3	w_3x_3	y_3	w_3y_3
	$\Sigma w = W$		$\Sigma wx = W\bar{x}$		$\Sigma wy = W\bar{y}$

whence $\bar{x} = \dfrac{W\bar{x}}{W}$ and $\bar{y} = \dfrac{W\bar{y}}{W}$

Determination of position of Centroid

Part	*Area*	*Distance of Centroid from YY*	*Moment about YY*	*Distance of Centroid from XX*	*Moment about XX*
1	a_1	x_1	a_1x_1	y_1	a_1y_1
2	a_2	x_2	a_2x_2	y_2	a_2y_2
3	a_3	x_3	a_3x_3	y_3	a_3y_3
	$\Sigma a = A$		$\Sigma ax = A\bar{x}$		$\Sigma ay = A\bar{y}$

whence $\bar{x} = \dfrac{A\bar{x}}{A}$ and $\bar{y} = \dfrac{A\bar{y}}{A}$

Note: It should not be overlooked that equations given on p. 198, can be re-written in the following forms:

$$w_1x_1 = W\bar{x} - w_2x_2$$

i.e.
$$x_1 = \frac{W\bar{x} - w_2x_2}{w_1}$$

which allows the determination of the position of the centre of gravity of a solid from which a piece of material (w_2) has been removed. Also

$$a_1x_1 = A\bar{x} - a_2x_2$$

i.e.
$$x_1 = \frac{A\bar{x} - a_2x_2}{a_1}$$

for the determination of the position of the centroid of an area from which part (a_2) has been removed.

Example

A steel forging is made up of three parts, each cylindrical in form, and of dimensions shown in figure 11.4. If the weight of steel is 0·28 lbf/in³ determine the position of the centre of gravity from the left hand end of the forging.

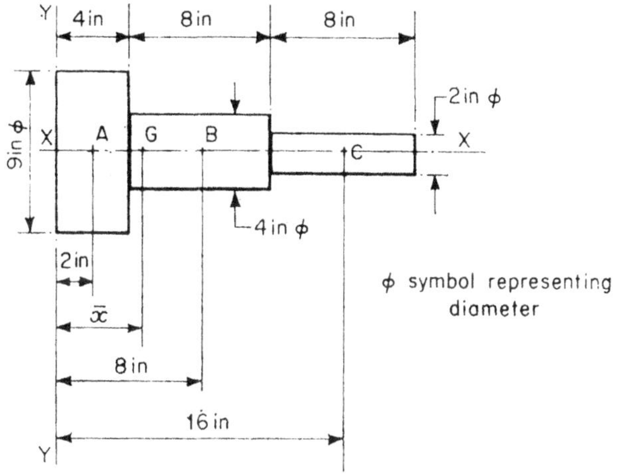

Figure 11.4. A steel forging

Solution

Because the three parts—marked A, B and C—are plain cylinders, their centres of gravity must be mid-way along their respective lengths. The distances of these from the left hand end, indicated by the vertical line YY, have been added to figure 11.4 so as to clarify the solution.

$$\text{Weight of a steel cylinder} = 0{\cdot}28 \times \text{Volume}$$

$$= 0{\cdot}28 \times \frac{\pi}{4} \times d^2 l \text{ lbf if } d \text{ and } l \text{ are in inches}$$

$$= \text{Constant} \times d^2 l$$

Substituting in the equation

$$W\bar{x} = w_1 x_1 + w_2 x_2 + w_3 x_3$$

$$0{\cdot}28 \times \frac{\pi}{4} \times \Sigma d^2 l \times \bar{x} = 0{\cdot}28 \times \frac{\pi}{4} [(d_1^2 l_1 \times x_1) + (d_2^2 l_2 \times x_2) + (d_3^2 l_3 \times x_3)]$$

or, $$\Sigma d^2 l \times \bar{x} = d_1^2 l_1 \times x_1 + d_2^2 l_2 \times x_2 + d_3^2 l_3 \times x_3$$

showing that there is no necessity to make use of the specific weight of steel. Use the tabular form of solution taking care to note that the same unit of length is used throughout—in this case the inch.

Part	d^2l	x	$d^2l \times x$
A	324	2	648
B	128	8	1024
C	32	16	512
	484 $= \Sigma d^2l$		2184 $= \Sigma(d^2l \times x)$

$$\bar{x} = \frac{\Sigma(d^2l \times x)}{\Sigma d^2l} = \frac{2184}{484}$$

$\bar{x} = 4{\cdot}51$ in. from the left hand end.

The centre of gravity of the forging is, of course, on the symmetrical axis XX.

Example

A circular section shaft, 1·6 metres long and 0·12 metre in diameter, is made from material having a specific mass of 7,800 kilogramme mass per cubic metre. Flywheels weighing 450 and 750 kilogrammes are located 0·15 and 0·25 metre respectively from either end of the shaft. Calculate the position of the centre of gravity of the combined system.

Solution

$$\text{Mass of shaft} = \left(\frac{\pi}{4} \times (0{\cdot}12)^2 \times 1{\cdot}6\right) \text{m}^3 \times 7{,}800 \text{ kg/m}^3$$
$$= 141 \text{ kg}$$

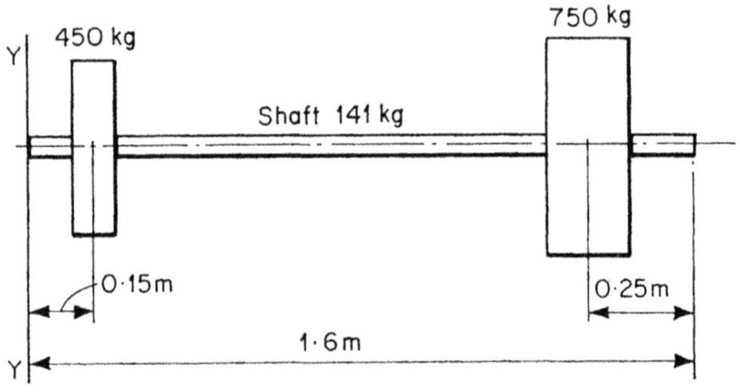

Figure 11.5

The shaft is uniform and therefore the mass may be considered, in subsequent calculation, to be concentrated at its mid-length position.

Use vertical YY (left hand extremity of shaft) as the zero for distance.

Mass	*Distance of C.G. from YY*	*Moment about YY*
450	0·15	67·5
141	0·80	112·8
750	1·35	1012·5
$\Sigma w = 1341$		$\Sigma wx = 1192{\cdot}8$

$$\bar{x} = \frac{\Sigma wx}{\Sigma w}$$

$$= \frac{1192{\cdot}8}{1341}$$

$$= 0{\cdot}89 \text{ metre from } YY$$

(*Note:* The arithmetic would have been reduced if moments had been taken about an axis through the centre of gravity of one of the flywheels.)

Example

A thin sheet 12 inches square has a circular hole 3 inches diameter cut in it. The centre of the hole is 5 inches from one edge and 8 inches from an adjacent edge of the sheet. Calculate the position of the centroid relative to these two edges.

Solution

Referring to figure 11.6 let $\bar{x}$ and $\bar{y}$ be the distances of the centroid from the adjacent edges YY and XX respectively.

$$\text{Area of square sheet} = 144 \text{ in}^2 = a_1, \text{ say}$$

$$\text{Area of circular hole} = \frac{\pi}{4} \times 3^2 = 7{\cdot}07 \text{ in}^2, = a_2, \text{ say}$$

The solution can be obtained conveniently by using the tabular method.

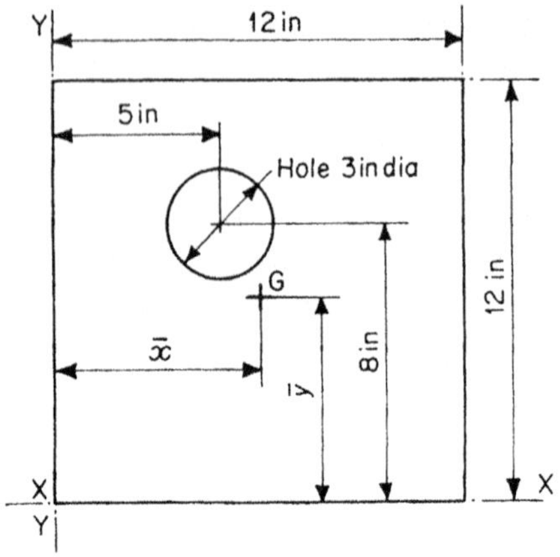

Figure 11.6 Square sheet with circular hole.

Part	*Area*	*Distance of Centroid from YY*	*Moment about YY*	*Distance of Centroid from XX*	*Moment about XX*
a_1	144	6	864	6	864
a_2	7·07	5	35·35	8	56·56
	136·93 $= a_1 - a_2$		828·65 $= a_1x_1 - a_2x_2$		807·44 $= a_1y_1 - a_2y_2$

$$\text{whence } \bar{x} = \frac{828{\cdot}65}{136{\cdot}93} = 6{\cdot}05 \text{ in.}$$

$$\text{and } \quad \bar{y} = \frac{807{\cdot}44}{136{\cdot}93} = 5{\cdot}90 \text{ in.}$$

Example

An unequal angle (L-section) has the following dimensions:

$$15 \text{ cm} \times 10 \text{ cm} \times 1{\cdot}25 \text{ cm}$$

Ignoring the radii at root and toe calculate the position of the centroid of the section.

Solution

Consider the section as made up of two rectangles:

$$a_1 = 15 \text{ cm} \times 1{\cdot}25 \text{ cm} = 18{\cdot}75 \text{ cm}^2$$
$$a_2 = 8{\cdot}75 \text{ cm} \times 1{\cdot}25 \text{ cm} = 10{\cdot}94 \text{ cm}^2$$
$$A = a_1 + a_2 = 29{\cdot}69 \text{ cm}^2$$

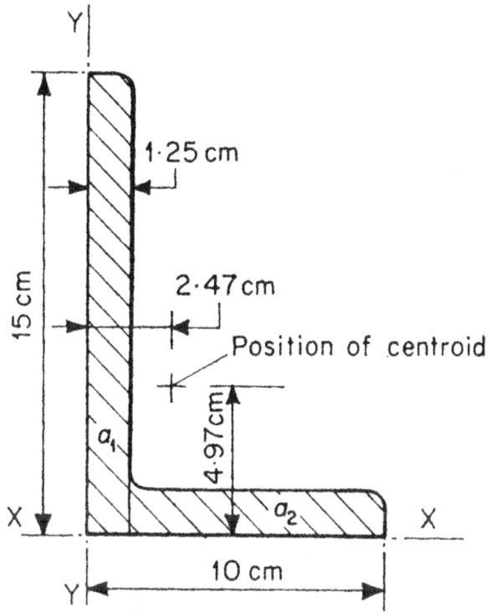

Figure 11.7

Part	*Area*	*Distance of Centroid from YY*	*Moment about YY*	*Distance of Centroid from XX*	*Moment about XX*
a_1	18·75	0·625	11·70	7·50	140·63
a_2	10·94	5·625	61·50	0·625	6·84
$\Sigma a = 29{\cdot}69$		$\Sigma ax = 73{\cdot}20$		$\Sigma ay = 147{\cdot}47$	

$$\bar{x} = \frac{\Sigma ax}{\Sigma a} = \frac{73{\cdot}20}{29{\cdot}69} = 2{\cdot}47 \text{ cm}$$

$$\bar{y} = \frac{\Sigma ay}{\Sigma a} = \frac{147{\cdot}47}{29{\cdot}69} = 4{\cdot}97 \text{ cm}$$

These dimensions, which define the position of the centroid have been added in detail to the section shown in figure 11.7. Note the centroid lies outside the perimeter of the section. As a further exercise, work through this example considering the section made of rectangle 15 cm × 10 cm *minus* rectangle 13·75 cm × 8·75 cm.

Summary

Centre of gravity

Each particle of a body is attracted towards the Earth's centre by the force of gravity. The resultant of all these gravitational pulls on the body is a measure of its weight.

Whichever way the body is tilted there is one point in it which lies on the line of action of the resultant gravitational force. This point, through which the weight of the body may be taken to act, is called the 'centre of gravity' of the body. If w is the weight of each elementary particle and x its distance from a reference axis then, in symbols,

Distance from reference axis to centre of gravity,

$$\bar{x} = \frac{\Sigma wx}{\Sigma w}$$

Similarly, distance from a second axis may be expressed as

$$\bar{y} = \frac{\Sigma wy}{\Sigma w}$$

For a symmetrical body the position of the centre of gravity may be obvious by inspection.

Centroid

In thin flat sheets the masses of individual elements are proportional to the surface areas of those elements. w, the weight of an element, is then proportional to a, its area on the surface of the sheet. In a way similar to that of the previous paragraph,

$$\bar{x} = \frac{\Sigma ax}{\Sigma a} \quad \text{and} \quad \bar{y} = \frac{\Sigma ay}{\Sigma a}$$

where $\bar{x}$ and $\bar{y}$ are the distances of the centre of area or *centroid* of the surface from the reference axes.

The principle may be applied to determine the position of the centroid of a plane section of a body.

Tabular solutions

The solution of problems on the positions of centres of gravity, or of centroids, may be carried out conveniently by arranging the work in tabular form. Examples of this method are worked out in the text.

Exercises

1. A metal plate with a rectangular opening is shown. If the thickness of the metal is uniform, determine the position of the centroid relative to the bottom and right hand edges.

2. A flat steel plate is 24 cm square. How far will its centroid be moved if two holes are drilled as shown?

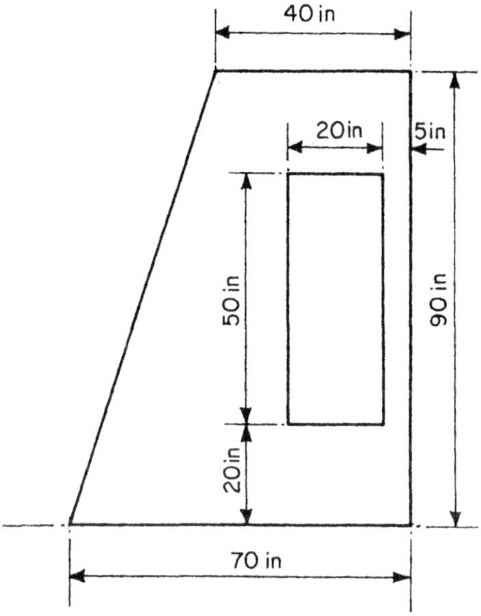

Figure 11.A1

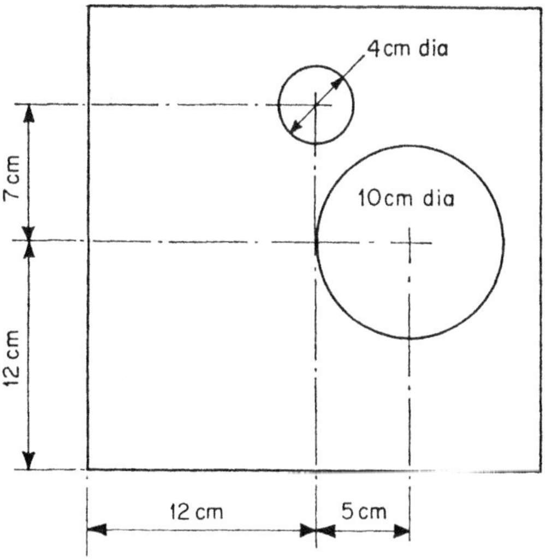

Figure 11.A2

3. A circular plate, 2 ft diameter, has a hole 4 in. diameter cut in it as shown in the figure. Find the position of its centroid relative to the centre of the plate.

4. Define the centroid of a section. Find the position of the centroid of the section shown relative to the bottom and right hand edges.

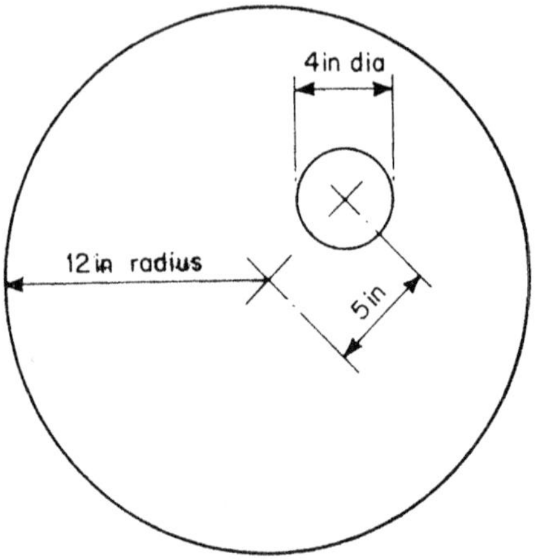

Figure 11.A3

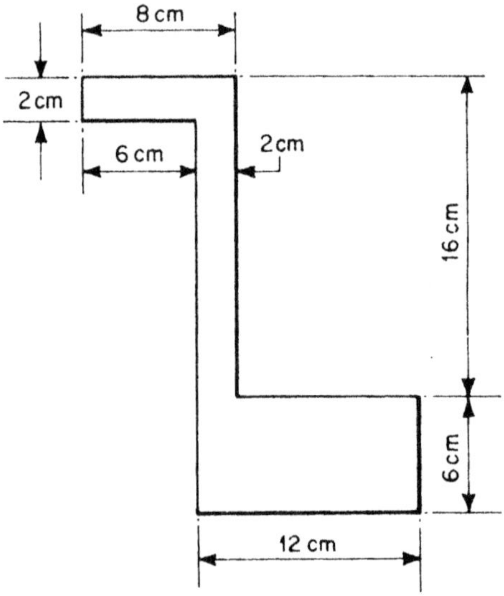

Figure 11.A4

5. A circular bar 15 in. long has 8 in. of its length turned to 2 in. diameter, and the remainder turned down to $1\frac{1}{2}$ in. diameter. Calculate the position of its centre of gravity from the large end.

6. A cylindrical casting of 30 cm diameter has a hole 7·5 cm diameter drilled in it. If the axes of the hole and the casting are parallel and 5 cm apart, calculate the distance of the centre of gravity of the remainder from the geometrical axis of the cylinder.

7. A steel forging takes the form of four concentric cylinders placed end to end. Particulars of the cylinders are as follows:

(*a*) 2 in. diameter, 4 in. long
(*b*) 10 in. diameter, $3\frac{1}{2}$ in. long
(*c*) 4 in. diameter, 10 in. long
(*d*) 2 in. diameter, 6 in. long.

If the centre of gravity of each cylinder lies on the central axis, and is mid-way along its length, calculate the distance of the C.G. of the forging from one end.

8. A solid piece of steel consists of a cylindrical portion 36 cm long and a right cone of height 30 cm the base of the cone coinciding with one end of the cylinder.

Given that the C.G. of a cone is at a quarter of its height from the base, calculate the position of the C.G. of the solid from the cylindrical end.

9. The ends of a frustum of a cone are circles of diameter 12 in. and 5 in. respectively, and the distance between them is 6 in. Given that the C.G. of a right cone is at a quarter of its height from the base, calculate the distance of the C.G. of the frustum from its larger end.

10. An iron bar of small diameter is bent through a right angle so that each leg is 2 metres long. Find the position of its C.G. relative to one end.

11. A bearer bar has the cross-section shown in the figure. Determine the position of the centroid of the section relative to (*a*) AB and (*b*) AC. (N.C.T.E.C.)

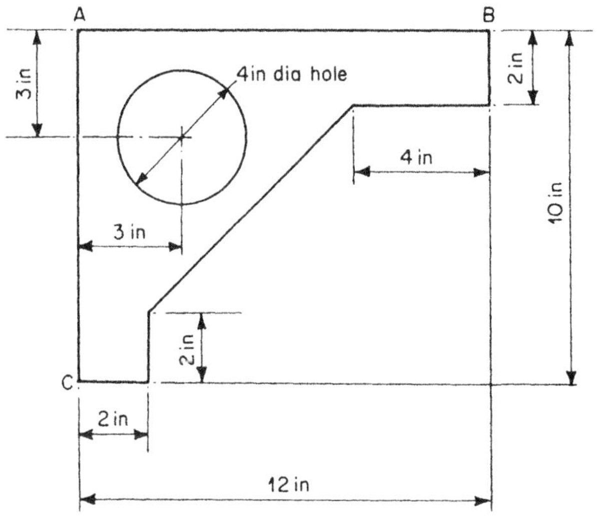

Figure 11.A11

12. A section of rail may be considered as having a top flange 7·24 cm wide and 3·88 cm deep, a bottom flange 10·30 cm wide and 2·54 cm deep and a web 2·16 cm thick and 7·86 cm deep, the overall depth of the section being 14·28 cm. The section is symmetrical about a vertical centre line. Calculate the height of the centroid.

12

The laws of a perfect gas

If a gas expands or contracts, certain changes usually occur to three of its physical properties. These properties are its volume, its pressure, and its temperature. The effect of changes in one or more of these physical properties can often be stated in terms of an algebraic expression, which we shall call a law of expansion or a law of compression. If a gas follows such laws completely and remains in a gaseous form throughout the change, it is said to be a perfect gas. Air behaves very nearly like a perfect gas.

Let us first of all consider what we mean by these physical properties.

Volume (symbol V)

The volume of a given weight of gas is measured in cubic feet (or in cubic metres in the Metric System). The numerical value of the volume is dependent on the pressure and temperature of the gas.

Pressure (symbol p)

Pressure is defined as the force per unit area. It can be, and usually is, measured directly in pounds per square inch, written lbf/in^2. (Newtons per square metre, written N/m^2 in the Metric System.)

In practice, pressures are measured above or below atmospheric pressure, and these pressures are either read off the dial of a Bourdon type pressure gauge, or determined from a manometer which is a form of U tube containing mercury or water.

Atmospheric pressure (symbol p_{at})
Atmospheric pressure means the pressure exerted at the Earth's surface by the weight of air above it. Because pressure is equal to the force

per unit area, the pressure on one square inch of the Earth's surface is equal to the weight of a column of air above it and supposedly contained in a tube of one square inch cross-sectional area and of a length equal to the height of the atmosphere above the Earth.

We can determine the pressure of the atmosphere at a particular place accurately by means of a standard Fortin's mercury barometer. In this barometer the pressure of the air is balanced by a vertical column of mercury, the scale being marked in inches and in centimetres.

Standard atmospheric pressure is usually accepted in British units as 14·7 lbf/in^2 or 30 inches of mercury. On the SI system, it is 101·3kN/m^2 or 760 mm of mercury.

Gauge pressure (symbol p_g)

The dial reading on a Bourdon type gauge gives the value of the pressure we want to measure, above or below the existing atmospheric pressure. We must, therefore, when we are recording a pressure reading from a gauge, determine the atmospheric pressure at the same time, and in the same place.

If the pressure inside the vessel containing a gas is less than atmospheric pressure (that is, a partial vacuum) the dial reading will be negative, and the pressure is then said to be a vacuum pressure, or just a vacuum. Vacuum pressures are normally measured in British units as inches of mercury and not in pounds per square inch, and the dial reading on a vacuum gauge will record the number of inches of mercury BELOW atmospheric pressure.

Absolute pressure (symbol p)

Absolute pressure is measured above zero pressure which corresponds to a perfect vacuum. Thus,

$$\text{Absolute pressure} = \text{Atmospheric pressure} + \text{Gauge pressure}$$

Atmospheric pressure here means the pressure of the atmosphere at the time and place of measurement of the gauge pressure.

If the gauge pressure is measured in inches of mercury, and the absolute pressure is required in pounds per square inch we need a conversion factor. Thus,

$$30 \text{ in. mercury} = 14{\cdot}7 \text{ lbf/in}^2$$

and in SI units,

$$76 \text{ cm mercury} = 101{\cdot}3 \text{ kN/m}^2$$

Figure 12.1 shows diagrammatically the relationship between the gauge pressure and the absolute pressure.

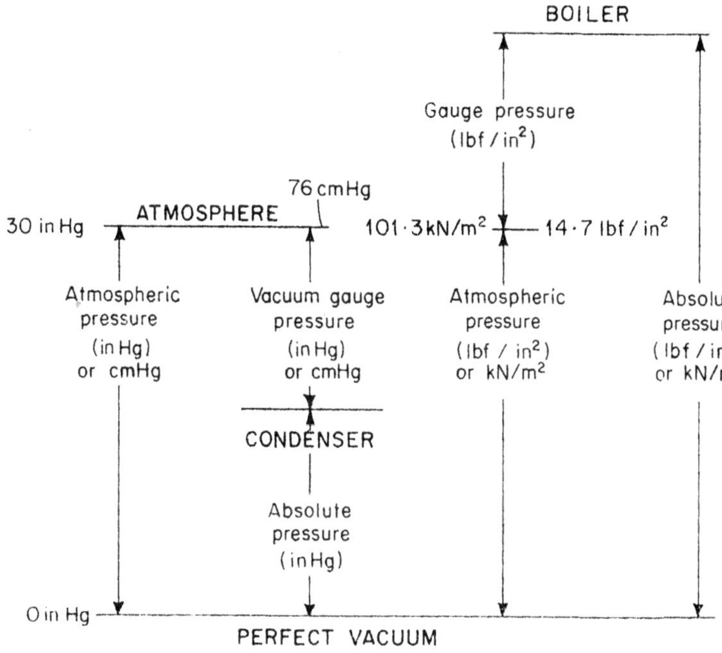

Figure 12.1 Absolute and gauge pressures.

Temperature (symbol t)

The temperature of a gas is its hotness or coldness expressed in terms of the amount of expansion or contraction of a gas, liquid, or a solid.

We can detect changes in temperature by means of a mercury-in-glass thermometer. Here, mercury expands and contracts by readable amounts in a fine capillary tube as the temperature rises and falls. A scale, made up of a number of equal divisions, provides the means of measuring the change in volume.

There are two temperature scales in common use: the Centigrade and the Fahrenheit scales. The latter is at present recommended for engineers, but it is necessary to be able to convert temperatures from one scale to the other. The temperature on the Centigrade scale is determined in degrees Celsius, and the temperature on the Fahrenheit scale in degrees Fahrenheit.

Mercury-in-glass thermometers use the freezing point and the boiling point of water at standard atmospheric pressure (14·7 lbf/in^2 or 101·3 kN/m^2) to provide datum lines on their scales. On the Centigrade scale, number 0 is assigned to the mark corresponding to the temperature at which water freezes, and number 100 to the mark corresponding to the temperature at which water boils. On the Fahrenheit scale numbers

32 and 212 are assigned to freezing and boiling temperature levels respectively.

Consequently, because each scale is divided into equal divisions, 100 divisions of the Centigrade scale are equal in length to $212 - 32 = 180$ divisions of the Fahrenheit scale.

That is, 1 division on the Centigrade scale is equal in length to 1·8 divisions on the Fahrenheit scale.

If we wish to convert the reading on one scale to a reading on the other scale, let

t_c = Temperature in degrees Celsius (measured on the Centigrade scale)

and t_f = Same temperature in degrees Fahrenheit

Then $$t_f = 32 + 1{\cdot}8t_c$$

or $$t_c = \frac{t_f - 32}{1{\cdot}8}$$

The relationship between the Centigrade and Fahrenheit scales is shown diagrammatically in figure 12.2.

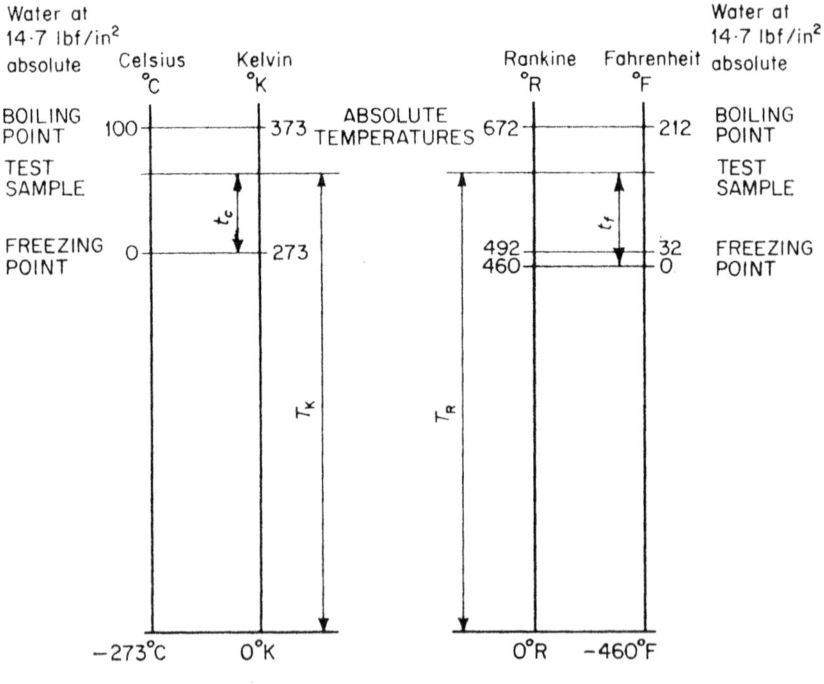

Figure 12.2 Temperature scales.

Absolute temperature (symbol T)
Any gas, on cooling, contracts in volume at a rate which can be determined experimentally. At some point, known as the absolute zero temperature, the gas should, if the contraction continued, occupy no volume. This temperature can only be deduced from the cooling rate, because the gas would liquefy, and probably solidify, before this low temperature could be reached. The point of absolute zero volume is estimated to occur at 273°C or 492°F BELOW the freezing point of water.

Absolute temperatures determined from the Celsius scale are stated in kelvins (K), the relationship being shown in figure 12.2.

Thus,

$$\text{Absolute temperature } (T_K) = \text{Thermometer reading } (t_c°) + 273$$

Absolute temperatures determined from the Fahrenheit scale are stated in degrees Rankine (°R). Thus,

$$\text{Absolute temperature } (T_R) = \text{Thermometer reading } (t_f°) + 460$$

The gas laws

The volume of a given weight of gas can be changed in any one of three ways:

1. by altering its pressure, keeping the temperature constant,
2. by altering its temperature, keeping the pressure constant,
3. by altering both its pressure and its temperature.

Each one of these ways leads us to an important gas law.

Boyle's Law

If the volume of a perfect gas is changed by altering its pressure, while keeping its temperature constant, the process is known as an isothermal change.

During this change of state, Boyle found, by experiment, that the volume of the gas varies inversely as its absolute pressure. Thus,

$$V \propto \frac{1}{p}$$

where V = Volume of the gas,

p = ABSOLUTE pressure of the gas.

This relationship is usually expressed in the form

$$pV = \text{Constant} \qquad 12.1$$

and is known as Boyle's Law.

If suffix 1 denotes the initial state and suffix 2 the final state of the gas, then

$$p_1V_1 = p_2V_2 = pV = \text{Constant}$$

which can be interpreted simply as meaning that, if we double the pressure of the gas, its volume is reduced to one half.

We can plot a graph of the change of volume. Its shape is a rectangular hyperbola, as shown in figure 12.3.

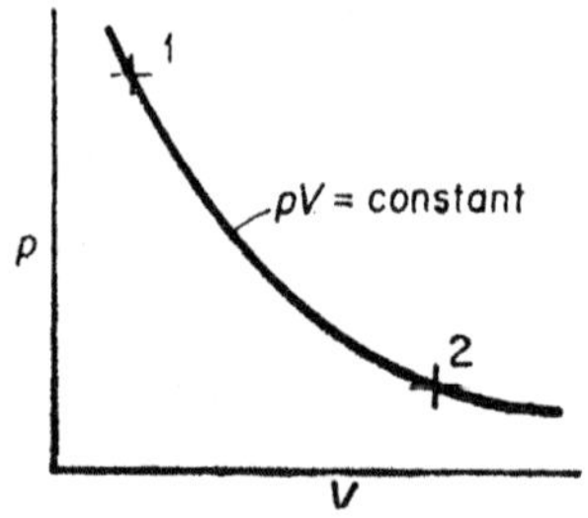

Figure 12.3 Boyle's law—pV = constant.

Charles' law

If the volume of a perfect gas is changed by altering the temperature, while keeping the pressure constant, its volume varies directly as its absolute temperature. Thus,

$$V \propto T$$

where

V = Volume of the gas

and

T = ABSOLUTE temperature in degrees Rankine or degrees kelvin

This relationship was discovered experimentally by Charles, and is usually expressed in the form

$$\frac{V}{T} = \text{Constant} \qquad 12.2$$

or, for two states,

$$\frac{V_1}{T_1} = \frac{V_2}{T_2} = \frac{V}{T} = \text{Constant}$$

where suffixes 1 and 2 refer to the initial and final states of the gas respectively.

The combined law

If the volume of a perfect gas is changed by altering both its pressure and its temperature, let the initial and final states be represented on a p-V diagram by the numbers 1 and 2.

If the points 1 and 2 in figure 12.4 do not lie on a known curve we

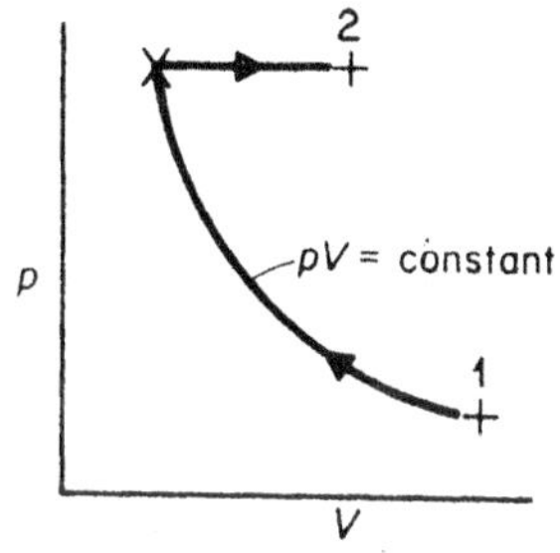

Figure 12.4 The combined law.

can move from state 1 to state 2 by a combination of the two processes we have just considered. Thus, suppose we move from state 1 to a point X at constant temperature on a pV = Constant curve, the point X being so selected that the gas here has the same pressure as if we had moved to state 2. We can then move from X to state 2 on a constant pressure line, which is horizontal.

From Boyle's Law

$$p_1V_1 = p_xV_x$$

and, from Charles' Law

$$\frac{V_x}{T_x} = \frac{V_2}{T_2}$$

Eliminating V_x from these equations

$$\frac{p_1V_1}{p_x} = \frac{V_2T_x}{T_2}$$

and, because $T_x = T_1$ on the pV = Constant curve, and $p_x = p_2$ on the constant pressure line,

$$\frac{p_1V_1}{p_2} = \frac{V_2T_1}{T_2}$$

or,

$$\frac{p_1V_1}{T_1} = \frac{p_2V_2}{T_2} \qquad 12.3$$

This is a most important expression and should be committed to memory. Both Boyle's and Charles' laws can be derived from it merely by applying the appropriate conditions—that is, constant temperature (or $T_1 = T_2$) for Boyle's Law and constant pressure (or $p_1 = p_2$) for Charles' Law.

The general expression is

$$\frac{pV}{T} = \text{Constant} \qquad 12.4$$

The expression 12.3 may be re-written in another form. Thus

$$\frac{V_1}{V_2} = \frac{p_2}{p_1} \times \frac{T_1}{T_2} \qquad 12.5$$

giving a relationship between the three properties in terms of ratios.

Consequently, we may use volumes and pressures in any units provided they are the same for both states. We must, however, ensure that

1. The pressures are absolute.
2. The temperatures are absolute.

The characteristic equation of a perfect gas

If p is the absolute pressure in lbf/in^2 and V is the volume of the gas, the general expression in equation 12.4 can be written

$$\frac{pV}{T} = \text{Constant}$$

If m is the weight of gas used, we may write

$$pV = mRT \qquad 12.6$$

where R is a constant for the gas.

Equation 12.6 is known as the Characteristic Equation. In this equation we must be careful to use pressures and volumes in their correct units.

Example

Calculate the value of the characteristic gas constant, R, for air. Normal atmospheric pressure is 1,013 mb, and the specific volume of air is 0·774 m^3/kgf at 0°C and one normal atmosphere pressure. Work in SI units. Note that 1 mb = 100 N/m^2.

Solution

From $PV = wRT$

$$R = \frac{PV}{wT}$$

$$= \frac{101{,}300\,\frac{\text{N}}{\text{m}^2} \times 0{\cdot}774\ \text{m}^3}{1\ \text{kgf} \times 273°\text{K}}$$

$$= 287\ \text{J/kg K}$$

$P = 101{,}300\,\dfrac{\text{N}}{\text{m}^2}$

$V = 0{\cdot}774\ \text{m}^3$

$w = 1\ \text{kgf}$

$T = 273°\text{K}$

Example

If 1 kg of air at 101·3 kN/m² absolute and 0°C occupies a volume of 0·775 m³, determine the temperature at which 20 kg of air will fill a volume of 6·2 m³ at a pressure of 170 kN/m² absolute.

Solution

In a question of this type determine first of all the volume occupied by 6·2 m³ of air under the standard, or known, conditions. Thus, at

101·3 kN/m² and 273 K 1 kg of air occupies 0·775 m³

Therefore, at

101·3 kN/m² and 273 K, 20 kg of air occupy $20 \times 0{\cdot}775 = 15{\cdot}5\ \text{m}^3$

We can now use the combined law to find the absolute temperature required, and hence the reading on the Celsius scale.

$$\frac{p_1V_1}{T_1} = \frac{p_2V_2}{T_2}$$

or,

$$T_2 = \frac{p_2}{p_1} \times \frac{V_2}{V_1} \times T_1$$

$$= \frac{170}{101{\cdot}3} \times \frac{6{\cdot}2}{15{\cdot}5} \times 273$$

$$= 183{\cdot}2\ \text{K}$$

$$\therefore\ t_2 = 183{\cdot}2 - 273$$

$$= -89{\cdot}8°\text{C}$$

$p_2 = 170\ \text{kN/m}^2$

$p_1 = 101{\cdot}3\ \text{kN/m}^2$

$V_2 = 6{\cdot}2\ \text{m}^3$

$V_1 = 0{\cdot}775\ \text{m}^3$

$T_1 = 273\ \text{K}$

Example

One cubic foot of air at 0°C and 14·7 lbf/in² absolute weighs 0·0807 lbf. Calculate the weight of air contained in a cylindrical vessel 3 ft in diameter and 7 ft long when the pressure of the air is 120 lbf/in² absolute and its temperature is 50°C.

Solution

$$\text{Volume of cylindrical vessel} = \frac{\pi}{4} \times 3^2 \times 7 = 49{\cdot}4 \text{ ft}^3$$

We shall first of all determine the volume of this air under the standard conditions. Thus

$$\frac{p_1 V_1}{T_1} = \frac{p_2 V_2}{T_2}$$

or,

$$V_1 = \frac{p_2}{p_1} \times \frac{T_1}{T_2} \times V_2$$

$$= \frac{120}{14{\cdot}7} \times \frac{273}{323} \times 49{\cdot}4$$

$$= 340 \text{ ft}^3$$

$p_2 = 120 \text{ lbf/in}^2$
$V_2 = 49{\cdot}4 \text{ ft}^3$
$T_2 = (273 + 50)°\text{K}$
$p_1 = 14{\cdot}7 \text{ lbf/in}^2$
$T_1 = (273 + 0)°\text{K}$

But, under the standard conditions (14·7 lbf/in² and 0°C)

1 ft³ of air weighs 0·0807 lbf

∴ 340 ft³ of air weigh 340 × 0·0807 = 27·4 lbf

Example

Two vessels A and B contain air and are connected by a pipe including a valve. When the valve is shut A contains 4 m³ at 14 b absolute and B contains 6 m³ at 3·5 b absolute. The temperature in each vessel being 15·5°C, determine the final pressure of the air if the valve is opened, the temperature remaining the same. Determine also the total weight of air contained by the vessels.

(R for air = 287 J/kg K) (U.L.C.I.)

Solution

$$\begin{matrix}\text{Weight of air} \\ \text{in A}\end{matrix} + \begin{matrix}\text{weight of air} \\ \text{in B}\end{matrix} = \begin{matrix}\text{total weight} \\ \text{of air}\end{matrix}$$

But, from equation 12.6,

$$pV = mRT$$

whence

$$m = \frac{pV}{RT}$$

so that, since

$$m_A + m_B = \text{total } m$$

$$\frac{p_A V_A}{R_A T_A} + \frac{p_B V_B}{R_B T_B} = \frac{p(V_A + V_B)}{RT}$$

Dividing through by $\frac{1}{RT}$, these being common,

$$P_A V_A + p_B V_B = pV$$

or,

$$(14 \times 4) + (3{\cdot}5 \times 6) = p \times 10$$

whence

$$p = 7{\cdot}7 \text{ b}$$

To find the total mass of air (m)

$$m = \frac{pV}{RT} \qquad p = 7{\cdot}7 \times 10^5\ \text{N/m}^2$$

$$V = 10\ \text{m}^3$$

$$= \frac{7{\cdot}7 \times 10^5 \times 10}{287 \times 288{\cdot}5} \qquad T = (273 + 15{\cdot}5)\ \text{K} = 288{\cdot}5\ \text{K}$$

$$= 93\ \text{kg}$$

Summary

The laws of a perfect gas

A gas which behaves in conformity with certain laws is said to be a 'Perfect Gas'. These laws refer to the states of volume, pressure and temperature of the gas, and their relationships.

Pressure

Standard atmospheric pressure. In the United Kingdom the standard atmospheric pressure is 30 in. of mercury, which is equivalent to a pressure of 14·7 lbf/in^2, or 101·3 kN/m^2.

Absolute pressure. This is obtained by adding to the recorded gauge pressure that due to the Earth's atmosphere, as taken from a barometer.

Temperature

Scales. The range between the boiling and freezing points of water at standard pressure is taken as 100 degrees on the Celsius scale, or 180 degrees on the Fahrenheit scale. Thus, 1 Celsius degree is equivalent to 1·8 Fahrenheit degrees.

Absolute temperature

This is obtained by adding 460°F to a Fahrenheit temperature, and giving the result in degrees Rankine (°R); or 273°C to a Celsius temperature, and giving the result in kelvins (K).

$$T_R = t_f + 460 \quad \text{and} \quad T_K = t_c + 273$$

Boyle's law

This can be stated: 'at constant temperature, the volume of a given weight of gas is inversely proportional to its absolute pressure'.

Symbolically, $$V \propto \frac{\text{Constant}}{p}$$

so that $$p_1V_1 = p_2V_2 = p_nV_n$$

Charles' law

This can be stated: 'at constant pressure, the volume of a given weight of gas is directly proportional to its absolute temperature'.

Symbolically, $V \propto \text{Constant} \times T$

so that
$$\frac{V_1}{T_1} = \frac{V_2}{T_2} = \frac{V_n}{T_n}$$

The combined gas law

The above laws can be combined to give the following relationship, when all three conditions are subject to variation:

$$\frac{p_1V_1}{T_1} = \frac{p_2V_2}{T_2} = \frac{p_nV_n}{T_n}$$

that is
$$\frac{pV}{T} = \text{Constant}$$

The characteristic equation of a perfect gas

For a given weight of gas at a pressure p lbf/in^2, volume V ft^3, and temperature $T°$ absolute, having a mass m kg.

$$\frac{pV}{T} = \text{Constant}$$

or,
$$pV = mRT$$

where R is a constant for unit mass of the gas.

Exercises

(All pressures are absolute unless otherwise stated.)

1. Two thermometers, one having a Celsius scale and the other a Fahrenheit scale, are placed in two different hot liquids. The temperature readings are 70°C and 122°F respectively. What will be the readings when the thermometers are interchanged?

2. An accurate Celsius thermometer reads 50°C when placed in a certain liquid. A Fahrenheit thermometer placed in the same liquid reads 121·8°F. Determine the error in the scale on the Fahrenheit thermometer.

3. Determine the temperature at which the scale readings on Celsius and Fahrenheit thermometers are the same.

4. If 10 ft^3 of air at a temperature of 20°C are heated at constant pressure until the volume is doubled, what will be the final temperature?

5. A cylinder contains 1 m^3 of air at 150°C and a pressure of 750 mm of mercury. Find its volume at 20°C when the pressure is 2000 mm of mercury.

6. 1 ft^3 of air at 450 lbf/in^2 and 120°F expands to double its volume when the pressure is reduced to 100 lbf/in^2. Calculate the change in temperature.

7. Given that 0·45 kg of air at 0°C and a pressure of 750 mm of mercury has a volume of 0·35 m^3, determine the temperature at which 4·5 kg of air at a pressure of 1,000 mm of mercury fills a vessel of capacity 1·25 m^3.

8. An air receiver has a capacity of 120 ft^3. It contains air at a pressure of 150 lbf/in^2 and a temperature of 25°C.

If 1 ft^3 of air at 0°C and 14·7 lbf/in^2 weighs 0·0807 lbf, calculate the weight of air in the receiver.

9. A cycle tyre contains 1000 cm^3 of air at a pressure of 3 atmospheres. If the tyre was originally completely flat and was inflated by a pump of 250 cm^3 capacity, how many strokes of the pump were necessary to inflate it, assuming the temperature remained constant during the operation.

10. What do you understand by the terms "gauge pressure" and "absolute pressure"?

A compressor running at 1200 rev/min draws in 3300 mm^3 of air at 1 b and 10°C at each revolution, and pumps it into a container of volume 0·42 m^3. How long does the compressor run in order to increase the pressure in the container from 6 b gauge to 16 b gauge if the temperature of the air in the container remains constant at 20°C?

11. Explain what is meant by a "perfect gas".

An air compressor is used to compress 7 m^3 of air at a pressure of 1 b into a receiver having a capacity of 0·55 m^3. If the temperature is kept constant, determine the final pressure in the receiver.

12. Write down the characteristic equation for a perfect gas, and explain the meaning of each of the symbols you use.

If 1 ft^3 of air at 14·7 lbf/in^2 and 32°F weighs 0·0807 lbf, calculate the gas constant for air in (*a*) Celsius units, (*b*) Fahrenheit units.

13. A vessel contains 5 kg of air at a pressure of 7 b absolute and a temperature of 30°C. The vessel is connected to another by a short pipe and the air exhausts into it. The final pressure in both vessels is 276 kN/m^2 absolute and temperature 10°C. If the pressure in the second vessel before the air entered was zero absolute find the volume of each vessel. (R for air = 287 J/kg K.) (U.L.C.I.)

13

Properties of water and steam

We have seen in Chapter 4 that energy is defined as the capacity for doing work, and that there are many kinds of energy. In this chapter we shall consider two other forms—known as heat energy and internal energy—with particular reference to the heating of water to produce steam.

Heat is a form of energy which passes from one body to another when a temperature difference exists between them. Because temperature is a measure of the activity of the molecules within the body, heat is said to flow from the more active molecules to the less active molecules, or from the hotter to the cooler portion of the body.

Internal energy (symbol U)

Let us consider what happens when heat energy is applied to water in order to form steam. When heat is applied the molecules move about faster. Heat energy is thus stored in the liquid as extra molecular movement, and the increase in the energy of the water is shown as a rise in temperature. There will also be a small increase in volume, but it is so small that it can be neglected. The speeding up of the molecules is said to increase their 'internal energy'.

Internal energy is, in fact, a property of the water, because its value at a particular state is not dependent upon the process through which it has passed to reach the given state. We are not, however, very interested in the precise value of the internal energy, but we shall often require to know the value of the CHANGE in internal energy with a change in state.

Saturation temperature

At the surface of water which is being heated there are molecules which are trying to jump out of the liquid. As more heat energy is applied a condition is reached where the water molecules have sufficient internal energy to jump clear away from the surface and to form steam. The temperature then remains constant and we have reached what is known as the 'boiling point' of water. Any further increase of heat energy merely shoots off more water molecules, but does not raise the temperature of the liquid. This temperature cannot change, unless the pressure changes, provided that both water and steam are present. In other words, for any particular value of the air pressure there is a definite temperature at which the water will boil and change into steam. This temperature is known as the 'saturation temperature'. It is important to note that when steam is being formed it can occur at only one temperature for a given value of the air pressure.

Effect of pressure

The boiling point, or the saturation temperature, of water falls as the pressure of air at its surface is reduced. Clearly the saturation temperature must, therefore, rise as the air pressure is increased.

Saturated steam

When water is boiling and steam is being formed at constant temperature the steam is said to be saturated. Its physical condition is such that on the slightest increase in pressure or decrease in temperature some of its molecules, which are in the form of a vapour, will change back into water.

Wet steam

Wet steam is a mixture of steam and water particles at the saturation temperature corresponding to the pressure in the vessel containing the steam. Steam is always wet if it is close to water.

Dry saturated steam

When there is no longer any water present, either in the steam, or as a liquid, and the temperature of the steam corresponds to the saturation temperature, the steam is said to be dry and saturated.

Superheated steam

If further heat energy is added to dry saturated steam without changing the pressure, the molecules will become more active and the temperature will rise. The steam is then said to be "superheated". Unlike saturated steam, superheated steam is defined by stating both the pressure and the temperature.

The difference between the actual steam temperature and the saturation temperature corresponding to the pressure in the container is called "the degrees of superheat".

Volume of steam

If water is heated and converted into steam at, or about, atmospheric pressure, its volume will increase to many hundred times the volume it occupied as water. If the pressure is increased, its volume will decrease, but even at a pressure of 20 bars its volume will still be about 100 times that of water of the same weight.

Steam which is superheated occupies even more space than it did in the saturated state.

Enthalpy (symbol H)

When water flows from one position to another in a pipe, energy is expended in pushing the column of liquid, which lies ahead, along the pipe. Work is done on the liquid in causing it to travel along the pipe at a given velocity. We can regard this work as a form of energy which moves along with the liquid and is only present when motion is taking place, and while there is a continuous supply of liquid under pressure. This work done is known as 'flow work'.

Enthalpy is defined as the sum of the internal energy and the flow work, and is another property of the liquid. The value of the enthalpy depends greatly on the value of the internal energy, but, again, we are usually only interested in the value of a CHANGE in enthalpy.

Specific heat (symbol c)

Specific heat is a ratio. When we speak of the specific heat of any substance we mean the quantity of heat required to raise the temperature of a given weight of that substance divided by the quantity of heat required to raise the temperature of the same weight of water by an equal amount under exactly the same conditions.

Because water requires much more heat to raise its temperature than do most other substances, the specific heats of water under certain conditions is said to be unity, and the specific heat of nearly all other substances have a value less than unity. It should be noted that the specific heat of superheated steam at constant pressure is about half that of water at atmospheric pressure.

Quantity of heat energy (symbol Q)
When we heat water to form steam we nearly always do so at constant pressure. Under these conditions we can say that

$$\left.\begin{array}{c}\text{Quantity of heat supplied at constant}\\ \text{pressure}\\ \text{or}\\ \text{Heat flow to the water}\end{array}\right\} = \text{Increase in enthalpy}$$

The quantity of heat required to raise the temperature of any substance can be obtained approximately from the following expression:

$$Q = w \times c \times (t_2 - t_1) \qquad 13.1$$

where w = Weight of substance being heated,

c = Specific heat of the substance,

$(t_2 - t_1)$ = Temperature rise.

The units of 'heat quantity' are based on the heat required to raise the temperature of unit weight of water by 1 degree. Thus, 1 Celsius Heat Unit (abbreviated to 1 Chu) is the quantity of heat required to raise the temperature of 1 lbf of water by 1 degree Celsius. Also, 1 British Thermal Unit (or 1 Btu) is the quantity of heat required to raise the temperature of 1 lbf of water by 1 degree Fahrenheit. The MKS unit of corresponding magnitude is the kilogramme calorie, which is the amount of heat required to raise the temperature of 1 kilogramme of water 1 Celsius degree, and is equal to 4·1868 kJ.

Because 1 division on the Centigrade scale is equal in length to 1·8 divisions on the Fahrenheit scale,

1 Chu must be equal to 1·8 Btu

Steam tables

The quantities of heat energy, in terms of either Joules or Chu's, necessary to raise the temperature of unit weight of water or steam at various values of constant pressure are frequently required by engineers. If the specific heat of steam were constant at different pressures there would be no difficulty in calculating the quantity of heat necessary by the use of equation 13.1. Steam, however, does not behave, when it is wet, as a perfect gas, although it behaves very nearly so when it is superheated. Consequently, steam tables have been compiled giving the quantity of heat units necessary to raise the temperature of steam at different pressures. The data are based on a number of experiments conducted over long periods of time and take into account the variation of the specific heat of water and steam with pressure.

The tables are arranged in vertical columns, the first column covering a wide range of pressures at small intervals. Opposite each pressure is listed the saturation temperature and the quantities of heat required to raise the temperature of unit weight of water from freezing point to its boiling point, and then to convert it into dry saturated steam at constant pressure. These heat quantities are known as ENTHALPIES.

Note very carefully that the datum line for temperature is the freezing point of water, so that if we start off with water at any other temperature we must make appropriate adjustments to the values of heat energy which we obtain from the steam tables.

Because the values of heat quantities are given in terms of a unit weight of water or steam, they are known as SPECIFIC ENTHALPIES.

Specific enthalpy of water (symbol h_f)

The specific enthalpy of water is the quantity of heat required to raise the temperature of unit mass of water from freezing point to a higher temperature. Because no change of state is involved this quantity is often known as the 'sensible heat'.

Steam tables should be used to give accurate values of the enthalpy of water at its saturation temperature—that is, when it is boiling. At lower temperatures the specific enthalpy may be obtained with reasonable accuracy at pressures up to 200 lbf/in^2 from the expressions

$$h_f = (t_f - 32) \text{ Btu/lbf} \qquad 13.2$$

or,

$$h_f = t_c \text{ Chu/lbf, or kJ/kg} \qquad 13.3$$

where t is the actual temperature of the water in degrees Fahrenheit (t_f) or degrees Celsius (t_c).

Latent heat of steam (symbol h_{fg})

The latent heat of steam is the quantity of heat required to change its state from water to steam. The values in steam tables refer to the heat quantity required to change unit weight of saturated water (i.e. 1 kg on the SI system)—that is boiling water—into dry saturated steam without changing its temperature.

The value of the latent heat will depend on the pressure, decreasing as pressure is increased, and must be obtained from tables.

Specific enthalpy of dry saturated steam (symbol h_g)

The specific enthalpy of dry saturated steam is the quantity of heat required to raise the temperature of unit mass (1 kg on the SI system) of water from freezing point to its saturation temperature and then to evaporate it COMPLETELY at that temperature. That is,

$$h_g = h_f + h_{fg} \qquad 13.4$$

Dryness fraction (symbol x)

If the steam is wet it contains a quantity of water particles suspended in it, and therefore requires a further supply of heat energy before it reaches the state of dry saturated steam.

Suppose that, in the latent heat stage—that is, during the heating of water which is already boiling—only x times h_{fg} heat units (where x is less than unity) have been supplied per unit weight of water, whereas h_{fg} heat units are required for complete evaporation to form dry steam. Each unit mass of wet steam will contain x mass units of dry steam and $(1 - x)$ mass units of water particles suspended in it. The value

$$\frac{xh_{fg}}{h_{fg}} = x$$

is known as the dryness fraction of the steam, and is always quoted as a decimal fraction.

Specific enthalpy of wet steam $= h$, and

$$h = h_f + xh_{fg} \qquad 13.5$$

Specific enthalpy of superheated steam (no particular symbol)

The specific enthalpy of superheated steam is given in steam tables for various steam pressures and at various degrees of superheat. Steam tables, if available, should always be used, even though it may be necessary, because of the large intervals, to interpolate for both pressure and temperature.

Only at pressures at, or very near to atmospheric pressure, is it permissible to calculate a value for the specific enthalpy from the expression:

$$h = h_g + h_{fg} + 0{\cdot}48(t - t_{sat}) \qquad 13.6$$

where t = Actual steam temperature,
t_{sat} = Saturation temperature,
$0{\cdot}48$ = Specific heat of superheated steam at atmospheric pressure.

Specific volume of steam (symbol V)

The volume of unit mass of dry saturated steam is given in steam tables for various pressures. If the steam is wet, the volume of water present is small and may be neglected. Hence,

Total volume of unit weight of wet steam $= xV$

If the steam is superheated, the specific volume is not given in steam tables and must be calculated. We shall not, however, be required to make such a calculation in this course.

Example

How many heat units are required to produce 10 kg steam of dryness 0·5 at 7 b absolute from feed water at 20°C, given that at 7 b, $h_f = 697$ kJ and $h_{fg} = 2067$ kJ.

Solution

Enthalpy of 1 kg of wet steam at 7 b $= h_f + xh_{fg}$

from equation 13·5

$= 697 + 0{\cdot}5(2067)$
$= 697 + 1033{\cdot}5$
$= 1730{\cdot}5$ kJ

$\therefore$ Enthalpy of 10 kg of wet steam at 7 b = 17,305 kJ

Heat already in 10 kg of feed water at 20°C $= 10(20 - 0)$
$= 200$ k Cal.
$= 200 \times 4{\cdot}1868$ kJ
$= 837$ kJ

$\therefore$ Heat supplied to 10 kg of feed water to produce wet steam

$= 17{,}305 - 837$
$= 16{,}468$ kJ

Example

How many heat units are required to produce 5 kg of steam of dryness 0·9 at 8 b absolute from feed water at 21°C, given that at 8 b,

$$h_f = 721 \text{ kJ and } h_{fg} = 2048 \text{ kJ}$$

Solution

Enthalpy of 1 kg of wet steam at 8 b $= 721 + 0{\cdot}9(2048)$
$= 721 + 1843$
$= 2564$ kJ

$\therefore$ Enthalpy of 5 kg of wet steam at 8 b = 12,820 kJ

Heat already in 5 kg of feed water at 21°C $= 5 \times 21$
$= 105$ k Cal.

$\therefore$ Heat supplied to 5 kg of feed water to produce wet steam

$= 12{,}821 - (105 \times 4{\cdot}1868)$
$= 12{,}381$ kJ

Example

Steam at a pressure of 74 lbf/in^2 absolute and dryness 0·70 is heated at constant pressure until it is dry. Calculate the increase in volume of

5 lbf of steam, given that the specific volume of steam at 74 lbf/in^2 is 5·890 ft^3.

Solution

$$\text{Volume of 5 lbf of wet steam at 74 lbf/in}^2 = 5(xV)\ \text{ft}^3$$
$$= 5 \times 0{\cdot}70 \times 5{\cdot}890$$
$$= 20{\cdot}62\ \text{ft}^3$$
$$\text{Volume of 5 lbf of dry steam at 74 lbf/in}^2 = 5 \times 5{\cdot}890$$
$$= 29{\cdot}45\ \text{ft}^3$$
$$\text{Increase in volume} = 29{\cdot}45 - 20{\cdot}62$$
$$= 8{\cdot}83\ \text{ft}^3$$

STEAM TABLES

(a) SI units kJ/kg

Absolute Pressure (bars)	*Saturation Temp. (°C)*	*Specific Enthalpy of Saturated Steam (kJ/kg)*			*Specific Enthalpy of Superheated Steam (kJ/kg)*		
		Liquid (h_f)	*Latent (h_{fg})*	*Total (h_g)*	*423 K abs*	*473 K abs*	*523 K abs*
1	99·6	417	2258	2675	2777	2876	2975
2	120·2	505	2202	2707	2770	2871	2971
3	133·5	561	2164	2725	2762	2866	2968
4	143·6	605	2134	2739	2753	2862	2965
6	158·8	670	2087	2757	—	2851	2958
7	165	697	2067	2764	—	2846	2955
8	170·4	721	2048	2769	—	2840	2951
10	179·9	763	2015	2778	—	2829	2944

(*b*) Celsius Units (Chu/lb)

(or kilogramme calories/kg)

Absolute Pressure		*Saturation Temperature*	*Specific Enthalpy of Saturated Steam*		*Specific Enthalpy of Superheated Steam (Degrees Celsius Superheat)*			
(lbf/in^2)	*kN/m^2*	*(°C)*	*Liquid (h_f)*	*Latent (h_{fg})*	60	80	100	120
14·7	101·3	100·0	100·0	539·3	669·3	679·0	688·8	698·5
100	689·4	164·3	165·7	496·1	694·3	704·7	715·0	725·2
120	827 4	171 7	173·5	490·4	696·9	707·4	717·8	728·1
140	965·3	178·3	180·4	485·3	699·1	709·7	720·2	730·6
160	1103	184·2	186·6	480·6	701·0	711·7	722·3	732·7
180	1241	189·5	192·3	476·3	704·2	715·1	725·8	736·4

SI Abstracts by courtesy of the Controller of HMSO

Summary

Heat

Heat is a form of energy which flows from one body to another at a lower temperature.

Internal energy

The internal energy of water is due to its molecular movement, which is caused by heat. The temperature of the water is an index of the internal energy it contains.

Internal energy is denoted by the symbol U. We shall be particularly interested in changes in internal energy in this chapter.

Saturation temperature
For any particular air pressure there is a definite temperature at which water will boil and change into steam. This is known as the 'saturation temperature' for that pressure.

Saturation temperature will rise as the pressure is increased.

Saturated steam
When water is boiling and steam is being formed at constant temperature, the steam is said to be saturated.

Wet steam
This is steam containing water particles. It can only exist at the saturation temperature.

Dry steam
This is steam containing no water particles. When it is at the saturation temperature, it is said to be 'dry saturated'. Above that temperature, it is said to be 'superheated'.

Specific volume of steam (symbol V)
The volume of unit weight of dry steam is called its specific volume. It is obtainable directly from steam tables.

Enthalpy (symbol H)
Enthalpy is the sum of the internal energy of water and the flow work. This property used to be called 'Total Heat'.

Specific heat (symbol c)
This is the ratio of the quantity of heat required to raise the temperature of a substance by a given amount, to the quantity of heat required to raise the same weight of water by the same amount.

Quantity of heat energy (symbol Q)
This is given by the expression:

$$Q = w \times c \times (t_2 - t_1)$$

The symbols have the meanings denoted in this chapter.

The steam tables
The enthalpies quoted in steam tables relate to unit weight of water, and are called 'Specific Enthalpies'.

Specific enthalpy of water (h_f)
This is the quantity of heat required to raise the temperature of unit weight of water from freezing to a higher temperature. It is sometimes

called 'sensible heat' to distinguish it from the 'latent heat' which water absorbs during a change of state from water to steam.

Up to 8 b

$$h_f = (t_f - 32) \text{ Btu/lbf}$$

or,
$$h_f = t_c \text{ Chu/lbf}$$

or, (SI)
$$h_f = t_c \text{ kJ/kg}$$

Latent heat (h_{fg})
This is the quantity of heat required to change the state of unit weight of saturated water into dry saturated steam without change of temperature.

Specific enthalpy (h_g) *of dry saturated steam*
This is the quantity of heat required to raise the temperature of unit weight of water from freezing temperature to boiling (saturation) temperature, and then to evaporate it completely at that temperature. In symbols,

$$h_g = h_f + h_{fg}$$

Dryness fraction (x)
This is the ratio

$$\frac{xh_{fg}}{h_{fg}}$$

where x is the weight of dry steam in unit weight of wet steam.

$$\text{Specific enthalpy of wet steam} = h = h_f + xh_{fg}$$

Specific enthalpy of superheated steam
This is equal to the specific enthalpy of dry saturated steam at the stated pressure, plus the additional heat required to raise its temperature by the amount of the degrees of superheat. At pressures near atmospheric, it is possible to assume that the specific heat of steam at constant pressure is 0·48, and to obtain specific enthalpy from the expression:

$$h = h_g + h_{fg} + 0{\cdot}48(t - t_{sat})$$

The volume of unit mass of wet steam is given by the product:

Dryness fraction × Volume of unit mass of wet steam or xV

Exercises

(See Steam Tables at the end of Chapter 13.) All pressures are absolute.

1. Calculate the mass of water particles suspended in 100 kg of steam which is 0·87 dry at a pressure of 7 b.

2. Steam leaves a boiler 97 per cent dry at a pressure of 8 b. How many kJ are required to completely dry the steam?

3. The specific enthalpy of wet steam at a pressure of 110 lbf/in^2 was found to be 646 Chu's. If the specific enthalpy and latent heat of saturated water at this pressure are respectively 169·8 Chu and 493·2 Chu, calculate the dryness fraction of the steam.

4. A boiler produces steam 98 per cent dry at a pressure of 6 b from feed water at 16°C. Calculate

(*a*) the specific enthalpy of the steam leaving the boiler,

(*b*) the total number of heat units (kJ) supplied to each 1 kg of water.

5. Steam of dryness 0·90 enters an engine at a pressure of 100 lbf/in^2 and is exhausted at a pressure of 14·7 lbf/in^2. If the dryness of the steam at the exhaust is 0·20, calculate the drop in enthalpy of the steam in passing through the engine.

6. Determine the quantity of heat units required to change 5 kg of dry, saturated steam at a pressure of 6 b into superheated steam at 523 K abs, the pressure remaining constant during the process.

7. How many heat units are required to produce 4 lbf of steam at a pressure of 100 lbf/in^2 from feed water at 20°C when (*a*) the steam is dry and saturated, (*b*) the steam is 0·80 dry, (*c*) the steam has 105 deg. C superheat.

8. How many heat units are required to produce 5 lbf of steam at a pressure of 180 lbf/in^2 from feed water at 55°F when (*a*) the steam is dry and saturated, (*b*) the steam is 0·87 dry, (*c*) the steam has a temperature of 450°F?

9. Explain the terms 'dryness fraction' and 'enthalpy' as applied to steam.

Steam at a pressure of 7 b absolute is passed into a tank containing 120 kg of water at a temperature of 16°C. The final weight of the water in the tank is 128 kg, and its final temperature is 49°C. The weight of the tank is 22 kg and the specific heat of the tank material is 0·12.

Determine the initial dryness fraction of the steam. Take the following values for steam at 7 b absolute.

Sensible heat 697 kJ/kg
Latent heat 2067 kJ/kg (E.M.E.U.)

10. Explain the meaning of the following terms: 'wet steam', 'saturated steam', and 'superheated steam'.

Determine the quantity of heat required to generate 1 lbf of steam at constant pressure from feed water at 60°F, when

(*a*) the dryness fraction of the steam is 0·9

(*b*) the steam is saturated and dry

(*c*) the steam has a temperature of 400°F

Saturation temperature of steam	358·4°F
Sensible heat	330·6 Btu/lbf
Latent heat	864·5 Btu/lbf
Specific heat of superheated steam	0·5

(E.M.E.U.)

14

Steam plant

Because some of us tend to think that steam is a relic of the past, we would do well to remind ourselves occasionally that steam is used both for heat and for power in a great number of large and small factories in this country. The reason is that it has such outstanding qualities. Thus, it is produced from water which is normally in plentiful supply, and can be obtained at a low cost. Both water and steam are clean to use and are odourless. Steam has a very high heat content, and gives up a large proportion of that heat, in the form of latent heat, without any change of temperature. Further, it can be used over and over again.

In a steam producing plant there are two circuits:

1. The main circuit for producing steam.
2. The cooling water circuit.

1. The main circuit for steam

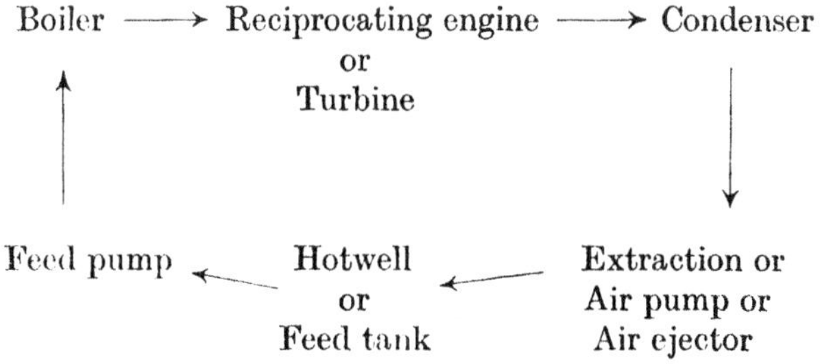

2. The cooling water circuit

Except when a jet condenser is used the two circuits are quite separate.

Condenser

Circulating pump ← River, Canal or Cooling tower

(Circulating pump → Condenser → River, Canal or Cooling tower → Circulating pump)

The main components of the producing plant are the boiler and the condenser. Of these there are many types.

Boilers

A boiler is a steam generator, the objects of which are:

1. To produce large quantities of heat by combustion of a fuel in the combustion chamber.
2. To transmit the heat of combustion to water and steam contained in the boiler drum.

The greater part of the heat energy in the fuel which is burnt in a boiler furnace is transmitted to the water and steam in the boiler, the rest being either carried up the chimney stack to waste, or lost by conduction, convection and radiation from the hot surfaces of the exterior of the boiler to the air surrounding the boiler.

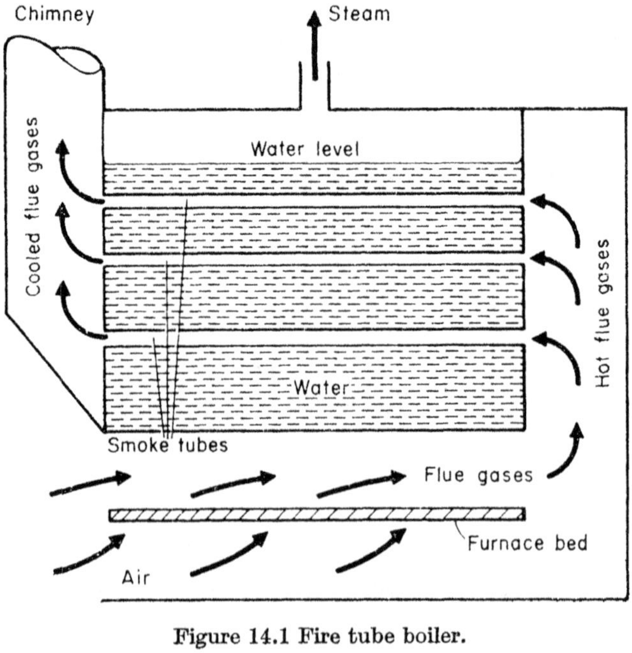

Figure 14.1 Fire tube boiler.

Steam boilers may be classified under two main headings:

1. Flue-tube, or fire-tube, boilers in which the hot gases flow through ducts or tubes which are surrounded with water.
2. Water-tube boilers in which the water circulates through tubes which are surrounded with hot gases.

Diagrammatic sketches of each type are given in figures 14.1 and 14.2.

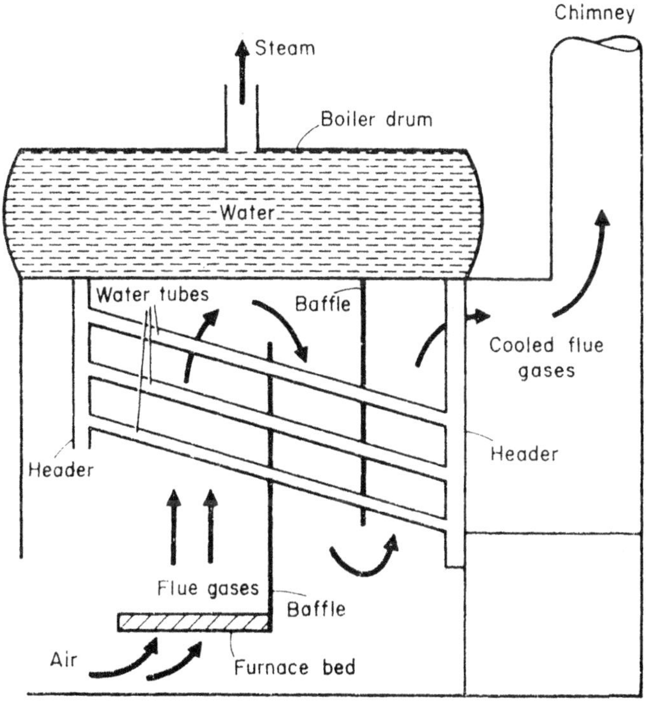

Figure 14.2 Water tube boiler.

Boiler efficiency

The thermal efficiency of a boiler is the ratio of the heat energy actually used in forming steam to the theoretical heat energy available in the fuel which is used to fire the boiler.

As with most efficiencies it is usual to express boiler efficiency as a percentage, thus

$$\text{Thermal efficiency of boiler} = \frac{\text{Output}}{\text{Input}} \times 100 \text{ per cent}$$

$$= \frac{\text{Heat energy used in forming steam}}{\text{Theoretical heat energy in fuel}} \times 100 \text{ per cent}$$

The output of a boiler is usually expressed in terms of the weight of water evaporated (or the weight of steam produced) per pound of fuel burnt. Sometimes it is expressed in terms of pounds of water evaporated per hour.

The heat energy used in forming the steam is calculated from the values of enthalpy given in steam tables. Thus, if

h = Specific enthalpy of steam at the pressure in the boiler,

and h_f = Specific enthalpy of the feed water pumped into the boiler.

Then,

Heat energy supplied to water to produce steam
$= (h - h_f)$ per unit mass

Note that, because the feed water is supplied at, or near to, atmospheric pressure, its specific enthalpy can be calculated in accordance with equations 13.2 and 13.3 of the previous chapter.

If W_s = Weight of steam produced. Then,

$$\text{Output} = W_s(h - h_f) \text{ energy units}$$

The input is calculated from the heating value of the fuel, known as its calorific value, and determined experimentally. Thus, if

$$W_f = \text{Weight of fuel consumed}$$

$$\text{Theoretical input} = W_f \times \text{Calorific value in energy units}$$

$$\text{Thermal efficiency of boiler} = \frac{W_s(h - h_f)}{W_f \times \text{Cal. value}} \times 100 \text{ per cent} \qquad 14.1$$

Example

A boiler burns coal of calorific value 31,400 kJ/kg and consumes 1585 kJ/h. Steam, with a dryness fraction of 0·95 is produced at a pressure of 11 b absolute, using feed water at a temperature of 27°C. If 14,500 kg of steam are generated an hour calculate the thermal efficiency of the boiler.

At 11 b absolute sensible heat of steam is 781 kJ/kg and the latent heat is 2000 kJ/kg.

Solution

$$\text{Heat input per hour} = 1585 \times 31{,}400 \text{ kJ}$$

$$\begin{aligned}\text{Heat output per hour} &= 14{,}500\,[781 + 0{\cdot}95\,(2{,}000)]\\ &= 14{,}500(781 + 1900)\\ &= 14{,}500 \times 2681 \text{ kJ}\end{aligned}$$

But the feed water already has a

$$\text{Heat content} = 14{,}500(27 - 0) \times 4{\cdot}1868 \text{ kJ}$$
$$= 391{,}500 \times 4{\cdot}187 \text{ kJ}$$
$$\therefore \text{Heat given to feed water per hour} = 14{,}500[2681 - 27 \times 4{\cdot}1868]$$
$$= 14{,}500 \times 2568 \text{ kJ}$$
$$\text{Thermal efficiency} = \frac{\text{Output}}{\text{Input}} \times 100 \text{ per cent}$$
$$\frac{14{,}500 \times 2568 \times 100}{1{,}585 \times 31{,}400}$$
$$= 75 \text{ per cent}$$

Condensers

Steam which has given up some of its heat in passing through the cylinder of a reciprocating engine can be exhausted (*a*) into the atmosphere as in a steam locomotive, (*b*) into another cylinder for further expansion, as in a compound marine engine, or (*c*) into a condenser where it is cooled and often used again.

The function of a condenser is to maintain a low pressure, so as to obtain the maximum possible energy from the steam and to secure a high efficiency. Consequently, all condensers maintain a vacuum pressure under normal working conditions.

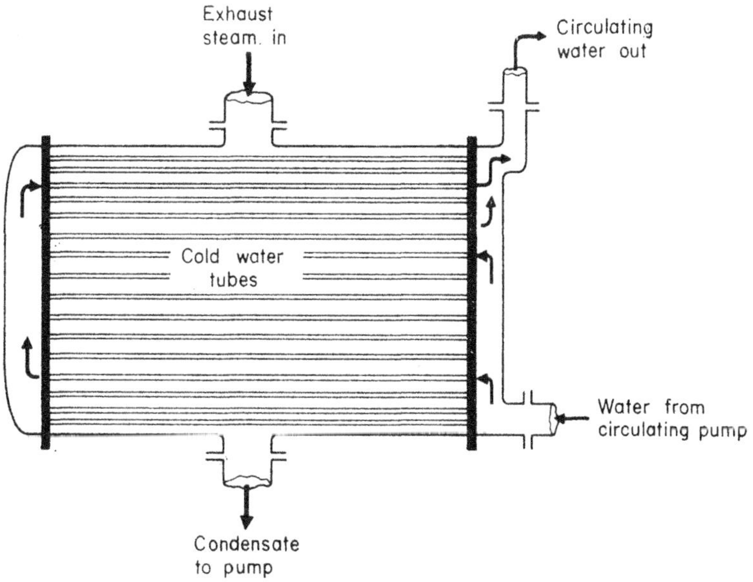

Figure 14.3 Surface condenser.

There are several types of condenser, the two most commonly used being the surface condenser and the jet condenser.

(*a*) *Surface condenser*
This type is most useful for marine work, but is also widely used in land installations. It is in popular demand when large quantities of inferior water are available. A typical surface condenser is shown in figure 14.3.

(*b*) *Jet condenser*
In this type of condenser the steam and the cooling water are in direct contact, the cooling water being sprayed into the incoming steam with the result that rapid and effective condensation occurs. A sketch of a typical jet condenser is given in figure 14.4.

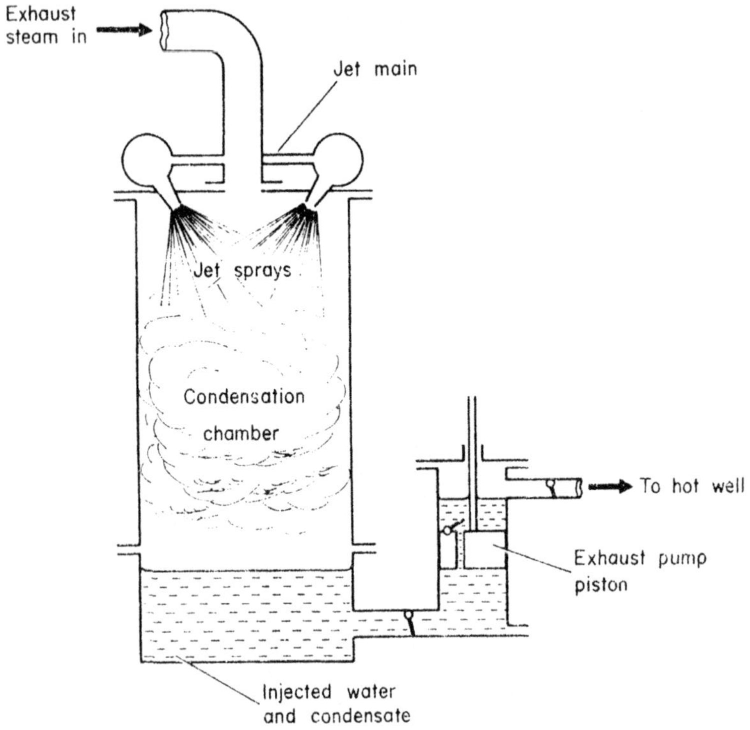

Figure 14.4 Jet condenser.

Heat exchange in a condenser

A condenser can be regarded as a heat exchanger, which is a means of transferring heat from a hot body to a cold body.

In a surface condenser the steam which enters it first of all loses its latent heat and then some of its sensible heat to the cooling water as it circulates through the condenser.

In a jet condenser the steam and the cooling water are mixed together, and so they emerge with a common temperature.

The heat exchange in both types of condenser can be determined by assuming that no losses occur, and, therefore, that

Heat lost by the steam = Heat gained by the cooling water

Example

In a test on a surface condenser, working at a pressure of 0·16 b absolute, 230 kg of steam of dryness 0·82 enter the condenser each hour, the condensate leaving at a temperature of 45°C. If the rise in temperature of the cooling water is 20 deg C calculate the weight of water circulating in tonnes per hour.

At 0·16 b abs.,

$$\text{Sensible heat} = 232 \text{ kJ/kg}$$

and

$$\text{Latent heat} = 2369 \text{ kJ/kg}$$

Solution

$$\begin{aligned}\text{Enthalpy of steam entering condenser} &= 232 + 0{\cdot}82(2369) \text{ kJ/kg}\\ &= 232 + 1943\\ &= 2175 \text{ kJ/kg}\\ \text{Enthalpy of condensate} &= (45 - 0) \times 4{\cdot}1868 \text{ kJ/kg}\\ \therefore \text{Heat lost by steam} &= 230(2175 - 188{\cdot}4)\\ &= 456{,}918 \text{ kJ/h}\end{aligned}$$

Let m = mass of circulating water in kg/h.

$$\text{Heat gained by circulating water} = m \times 20 \times 4{\cdot}1868 \text{ kJ/h}$$

$$= 83{\cdot}74m \text{ kJ/h}$$

But,

$$\text{Heat lost by steam} = \text{Heat gained by circulating water}$$

$$456{,}918 = 83{\cdot}74\, m$$

$$\therefore m = \frac{456{,}918}{83{\cdot}74} \text{ kg/h}$$

$$= 5{\cdot}456 \text{ t/h}$$

Summary

In a steam producing plant there are two circuits.

1. The main circuit for producing steam.
2. The cooling water circuit.

Boilers

These produce large quantities of heat by the combustion of a fuel in the combustion chamber. This heat is transmitted to the water and steam contained in the boiler.

In 'fire-tube' or 'flue-tube' boilers the hot gases flow through ducts or tubes which are surrounded with water.

In 'water-tube' boilers the water circulates through tubes which are surrounded with hot gases.

Boiler efficiency

$$\text{Thermal efficiency} = \frac{\text{Output}}{\text{Input}} \times 100 \text{ per cent}$$

$$= \frac{\text{Heat energy used in forming steam}}{\text{theoretical energy in fuel}} \times 100\%$$

In symbols,

$$\text{Thermal } \eta = \frac{W_s(h - h_f)}{W_f \times \text{calorific value}} \times 100 \text{ per cent}$$

Condensers

These are of two main types: surface condensers and jet condensers.

Exercises

1. A boiler generates 8 kg of steam per 1 kg of coal, the calorific value of which is 31,500 kJ/kg. Steam pressure at the boiler is 10 b and the dryness fraction of the steam is 0·92. If the feed water temperature is 16°C calculate the thermal efficiency of the boiler. (At 10 b the specific enthalpy of saturated water is 763 kJ/kg and the latent heat of steam is 2015 kJ/kg.)

2. A boiler generates 9·5 lbf of steam at a pressure of 140 lbf/in^2 and dryness fraction 0·92 for each 1 lbf of fuel burnt. The calorific value of the fuel is 13,500 Btu/lbf and the feed water temperature is 70°F. Calculate the thermal efficiency of the boiler. (At 140 lbf/in^2 the specific enthalpy of saturated water is 324·9 Btu/lbf, and the latent heat of steam is 869·1 Btu/lbf.)

3. Steam at a pressure of 2·6 lbf/in^2 and dryness fraction 0·87 enters a surface condenser at the rate of 5 lbf/s. The temperature of the condensate

is 97°F. If the rise in temperature of the cooling water is 27 deg. F calculate its weight in lbf/min. (At 2·6 lbf/in^2 the specific enthalpy of saturated water is 103·8 Btu/lbf, and the latent heat of steam is 1016·5 Btu/lbf.)

4. Steam at a pressure of 0·1 b and dryness fraction 0·91 enters a surface condenser. The temperatures of the cooling water at inlet and outlet are 28·8°C and 38°C respectively, and the temperature of the condensate is 43·3°C. If 13,590 kg of steam are condensed in one hour, calculate the number of litres of cooling water used per minute. (At 0·1 b the specific enthalpy of saturated water is 192 kJ/kg, and the latent heat of steam is 2392 kJ/kg.)

5. In a trial on a steam boiler the following results were obtained: steam per hour 1140 kg at 10 b absolute having a dryness fraction of 0·98. Feed water temperature 22°C. Coal per hour 220 kg, of calorific value 26,000 kJ/kg.

Calculate the boiler efficiency and also the volume of 1 kg of this steam. (*Note:* specific volume of steam at 10 b is 0·194 m^3/kg.)

(U.L.C.I.)

15

Internal combustion engines

In a steam engine the working substance—that is, steam—receives heat energy from a fuel burnt in a furnace, which may be sited at some considerable distance from the engine. Such an engine may be referred to as an external combustion engine.

Space and weight may be saved if the fuel is made to give up its heat energy to the working substance by burning inside the cylinder and the engine is then known as an internal combustion engine.

There are two main types of internal combustion (I.C.) engine:

1. reciprocating, or piston engines,
2. gas turbines.

Piston engines

In a piston-type I.C. engine the working substance is mainly air. A relatively small amount of fuel is burnt inside the engine cylinder and gives up its heat to the air, which, by expanding, causes the piston to move in the cylinder. This linear movement is then converted into rotary movement by means of the connecting rod and a crank on the output shaft.

The fuel which supplies the heat energy to the working substance may be petrol, gas or oil.

In any piston-type I.C. engine it is essential that arrangements be made for a cycle of four operations to occur. These are:

1. Fuel and air are supplied to the engine cylinder in the correct proportion.
2. Both the fuel and the air are compressed. This may be either before, or after, the two are mixed.

3. The compressed mixture is ignited, causing it to expand and produce mechanical movement to a piston.
4. The burnt products of combustion are removed from the cylinder to make way for a fresh charge.

This cycle of operations may occur while the piston completes four strokes (a stroke being the maximum displacement of the piston in the cylinder), in which case the engine is known as a 4-stroke engine, or the cycle may occur while the piston completes only two strokes, in which case the engine is known as a 2-stroke engine.

In a 4-stroke engine it is necessary to have an outlet valve and at least one inlet valve. In an engine which runs on gas it may be necessary to have separate fuel and air inlet valves.

In a 2-stroke engine there are usually no mechanically operated valves, but the piston, as it moves along the cylinder, covers or uncovers inlet and outlet ports.

Gas turbines

In its simplest form the gas turbine unit consists of an air compressor, a combustion chamber and a turbine. A line diagram of a unit is shown in figure 15.1.

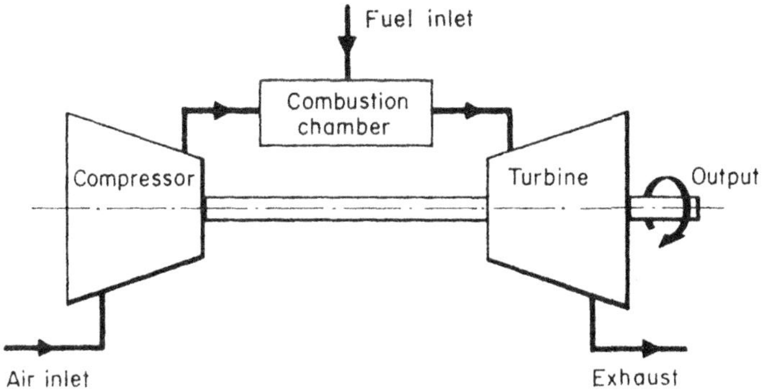

Figure 15.1 Gas turbine unit.

Air is drawn from the atmosphere into a rotary compressor where it is compressed to a high pressure before passing to a combustion chamber where it is heated by the internal combustion of the fuel. The hot gases at high pressure then pass to the turbine.

In the turbine the gases flow smoothly, but swiftly, in an axial direction through nozzles of varying cross-section designed to convert heat energy into kinetic energy. These nozzles are formed by the spaces between a large number of curved blades which are arranged round the periphery of a fixed disc. The gases emerge from these nozzles at very high speeds, and are directed on to a series of similar blades curved in the opposite direction and arranged round the periphery of a revolving wheel fixed to the shaft. A diagram showing the path of the gases through two sets of fixed blades and two sets of moving blades is given in figure 15.2.

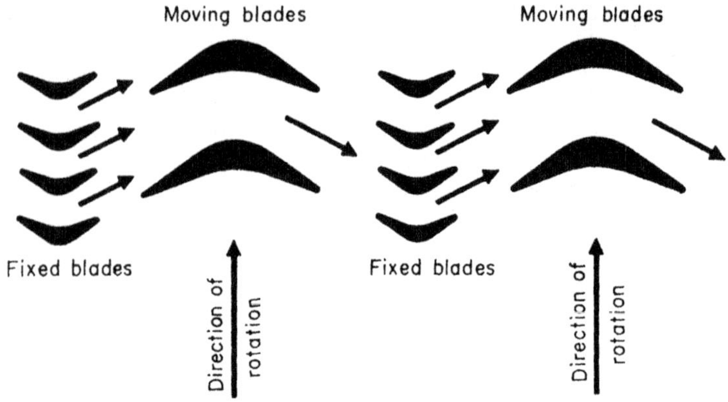

Figure 15.2 Path of gases through fixed and moving sets of blades.

The propulsive force on the moving wheels is produced by changing the direction, and hence the momentum, of the high speed gases.

A turbine usually has a number of revolving wheels, and between each pair of wheels—or rotors as they are called—will be a ring of fixed blades, or nozzles, curved in the opposite direction to the moving blades, and fixed to the casing of the turbine. The gases will, therefore, gain kinetic energy at the expense of heat energy as they flow in an axial direction through the turbine, and in doing so, cause the shaft to rotate and do work.

As they pass through the turbine the gases will expand, and so their volume at exit will be considerably greater than when they enter from the combustion chamber.

A rotary compressor may be regarded as a turbine acting in reverse. The compressor will also have a series of nozzles and a series of moving blades, but the function of the compressor is to increase the pressure of the air. In this process its volume is reduced and its temperature is increased.

The compressor is coupled mechanically to the turbine and has to be driven by it. Further, it absorbs the greater part of the energy at the output end of the shaft, and so only a small excess energy is available for doing useful external work.

16

Engine efficiencies

The amount of work done inside the cylinder of an engine may be determined with an indicator, which is an instrument designed for laboratory use to reproduce the relationship between pressure and volume of the working substance as the piston moves up and down the cylinder.

One type of indicator consists of a metal cylinder containing a small piston connected to a straight-line mechanism for reproducing in a vertical direction the pressure of the working substance in the engine cylinder, and a drum which is caused to oscillate about a vertical axis in phase with the motion of the engine piston.

A sectional view of a mechanical type of indicator is shown in figure 16.1.

Indicator diagram

A special paper, known as an indicator card, is wrapped round the drum of the indicator, and a stylus is fixed to the end of the long link in the straight-line mechanism. The working substance is admitted to the metal cylinder and there acts on the small piston, the motion of which is opposed by a calibrated spring. The movement of this small piston is conveyed through the linkage to the stylus which is caused to move vertically. When the drum is oscillating in phase with the engine piston the stylus is pressed on to the card. A diagram is produced to scale of pressure (vertically) against swept volume (horizontally). This diagram may be used to determine the horsepower developed by the engine.

The shape of the diagram varies with the type of engine. Thus an indicator used on a steam engine will produce a diagram of the shape shown in figure 16.2(*a*) whereas the diagram obtained from a diesel engine might have a shape such as that shown in figure 16.2(*b*).

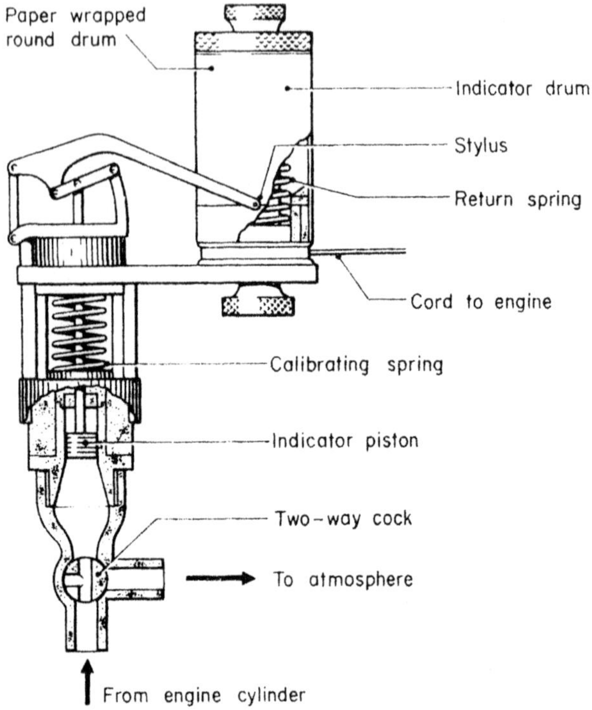

Figure 16.1 Sectional view of a mechanical type engine indicator.

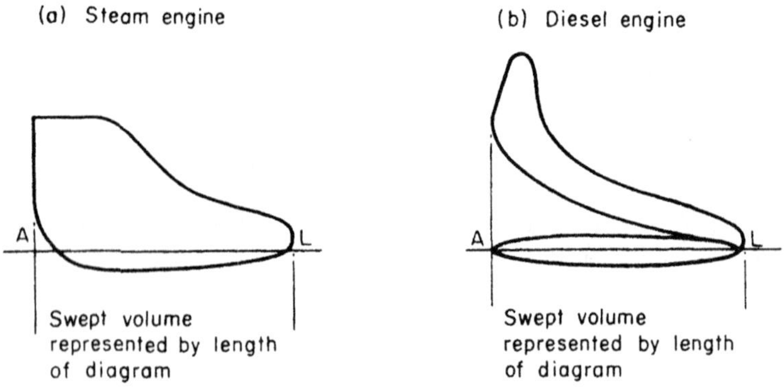

Figure 16.2 Engine indicator diagrams.

The horizontal line, A . . . L, is known as the atmospheric line, and indicates the atmospheric pressure at the time of the test. It can be drawn on the diagram by pressing the stylus against the drum while it is oscillating, and when the inlet valve to the indicator is closed to the engine but open to the atmosphere.

If two lines are drawn perpendicular to the atmospheric line so as to just touch the diagram at its two extremities, the distance between them is a measure of the swept volume of the engine piston. The 'swept volume' means the volume swept out by the engine piston during one stroke. It is, therefore, equal to the cross-sectional area of the cylinder multiplied by the stroke length, the stroke being twice the length of the crank.

The area of an indicator diagram may be determined in one of several ways. In practice it is found easiest to use a planimeter, which is an instrument for determining automatically the area enclosed when its stylus is guided round the perimeter of the diagram.

The mean height of the diagram in inches is equal to the area in square inches divided by the length in inches, and if the rate of the indicator spring is known, this mean height can be interpreted in terms of pressure. Usually the vertical movement of the indicator piston is known in terms of the rate of the indicator spring used—that is, one inch of vertical movement represents n lbf/in^2. On the SI system it would be calibrated in N/m^2.

Because 1 inch of vertical movement of the stylus represents n lbf/in^2.

Mean height of diagram represents

$$\frac{\text{Area of diagram}}{\text{Length of diagram}} \times n\,\text{kN/m}^2$$

This is an average pressure, which if it acted on the engine piston throughout its stroke would produce an area equal to that of the actual indicator diagram. This average pressure is known as the 'mean effective pressure' (symbol p_m), and is illustrated in figure 16.3.

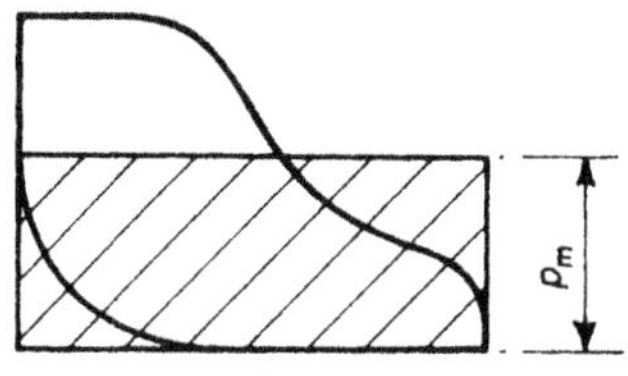

Figure 16.3 Mean effective pressure.

Note that the shaded area is equal to the area of the actual indicator diagram.

From this value of p_m the power developed in the cylinder can be calculated. Horsepower derived from an indicator card it is known as the 'indicated horsepower'.

Indicated horsepower (abbreviated to ihp)

Mean force on engine piston during its working stroke $= p_m A$ lbf

where $A =$ Area of piston (in²).

Work done by this force during a working stroke $= p_m A l$ ft lbf

where $l =$ Stroke of piston (ft)
$= 2 \times$ Length of crank.

Work done per minute $= p_m A l n'_m$ ft lbf

where $n'_m =$ Number of working strokes per minute.

$$\text{Indicated horsepower} = \frac{p_m A l n'_m}{33{,}000} \qquad 16.1$$

Note that A is determined in (inches)², but l is determined in feet.

On the SI system, Indicated power would be derived as $p_m A l n'_s$ watts, where p_m, A, l, would be expressed in newton and metre units.

Working strokes

A working stroke is another name for a power stroke, and indicates the frequency with which the working substance is admitted into the cylinder. Its value is usually determined in terms of revolutions per minute of the engine shaft which can be measured with a counter.

Thus, a 4 stroke engine admits the working substance once in every two revolutions of the engine crank shaft, whereas a 2 stroke engine admits the working substance once in every revolution of the crank shaft.

A steam engine is usually double-acting, which means that steam is admitted on each side of the piston, but in alternate strokes. Consequently, such an engine has two working strokes in every revolution of the crank shaft.

Brake power

The brake power of an engine is a measure of its output, and is determined in the laboratory by fitting a form of friction brake to a flywheel on the output shaft.

We have seen, in Chapter 7, how the value of the brake horsepower can be determined when using either a Rope Brake or a Prony Brake. With a rope brake, which is the more common of the two

$$\text{Brake horsepower (bhp)} = \frac{(W - S) \times 2\pi r n_m}{33{,}000}$$

where $(W - S)$ lbf $=$ Effective load on brake acting at a mean radius of r ft,

and $n_m =$ Speed of flywheel rev per min.

Note very carefully that the load is applied continuously to the rim of the flywheel, and therefore the brake horsepower (unlike the indicated horsepower) is INDEPENDENT of the number of working strokes, number of explosions, and number of cylinders.

For measurements on the SI system,

$$\text{Brake power} = (W - S) \times 2\pi r n_s \text{ watts}$$

Mechanical efficiency (symbol η_m, Greek letter eta)

The efficiency of an engine can be determined from the ratio

$$\frac{\text{Brake horsepower}}{\text{Indicated horsepower}} = \eta_m$$

Such an efficiency is called its 'mechanical efficiency'.

The difference between the brake horsepower and the indicated horsepower is known as the friction horsepower (fhp), and represents the loss in power from the input side of the engine—that is, produced in the cylinder—to the output shaft. Thus,

$$\eta_m = \frac{\text{bhp}}{\text{ihp}}$$

$$= \frac{\text{ihp} - \text{fhp}}{\text{ihp}}$$

$$= 1 - \frac{\text{fhp}}{\text{ihp}} \qquad 16.4$$

or, $$(\text{SI equivalent})\ \eta_m = 1 - \frac{\text{Friction power}}{\text{Indicated power}} \qquad 16.4$$

Thermal efficiency (symbol η_t)

The efficiency of an engine can also be expressed in terms of heat quantities. It is then known as a 'thermal efficiency'. Thus,

$$\eta_t = \frac{\text{Work done in terms of heat units per unit time}}{\text{Heating value of working substance in unit time}} \qquad 16.5$$

The work done may be stated as brake horsepower, in which case the efficiency is known as a 'brake thermal efficiency', or it may be stated as indicated horsepower, in which case the efficiency is known as an 'indicated thermal efficiency'.

The conversion of work done in ft lbf units to heat units is made by using Joule's Mechanical Equivalent of Heat. Thus,

1 Btu is equivalent to 778 ft lbf of work

or, 1 Chu is equivalent to 1400 ft lbf of work

In SI units,

$$1 \text{ kilogramme calorie} = 4186 \text{ joules}$$

Another name for the heating value of the working substance is its 'calorific value'. This value must be expressed in the SAME heat units as the work done in the numerator of the ratio (equation 16.5).

Example

During the trial of a 4 stroke diesel engine fitted with a rope brake the following observations were made:

Cylinder bore	165 mm
Stroke	254 mm
Mean effective pressure (from indicator card)	827 kN/m²
Speed	280 rev/min
Effective diameter of brake wheel	762 mm
Effective brake load	734 N
Fuel used	2·777 l
Calorific value of fuel	42,300 kJ/kg
Duration of test	1 hour

Calculate:

(*a*) ip, bp and mechanical efficiency,
(*b*) thermal efficiency on a bp basis.

Solution

$$\text{Working strokes/sec} = \frac{280}{2 \times 60} = \frac{140}{60} = n'_s$$

(*a*)

$$\text{ip} = plAn'_s$$

$$= 827 \times 0{\cdot}254 \times \frac{\pi}{4} \times (0{\cdot}165)^2 \times \frac{140}{60}$$

$$= 10{\cdot}48 \text{ kW}$$

$$\text{Braking torque} = 734 \times \frac{0{\cdot}762}{2} = T$$

$$\text{bp} = 2\pi nT = 2\pi \times \frac{280}{60} \times 734 \times \frac{0{\cdot}762}{2}$$

$$= 8200 \text{ N m/s} = 8{\cdot}2 \text{ kW}$$

$$\text{Mechanical efficiency } \eta_m = \frac{\text{bp}}{\text{ip}} = \frac{8{\cdot}2}{10{\cdot}48}$$

$$= 0{\cdot}782 \quad \text{or} \quad 78{\cdot}2 \text{ per cent}$$

(*b*) Brake thermal efficiency $= \dfrac{\text{Output in energy units per sec}}{\text{Heating value of fuel per sec}}$

$$\text{Heating value of fuel/sec} = \frac{42{,}300 \times 2{\cdot}777 \text{ kJ}}{3600}$$

$$= 32{\cdot}63 \text{ kJ}$$

$$\text{Brake thermal efficiency} = \frac{8{\cdot}2}{32{\cdot}63} \times 100 = 25{\cdot}13\%$$

Summary

Engine indicators; indicator diagrams

The pressure-volume relationship of gases expanding in a cylinder may be plotted on a card (called an 'indicator card') by using an automatic mechanism called an engine indicator.

The area of the diagram is measured by using a planimeter, and its mean height is then determined by dividing its area by its base length. This height, when multiplied by the indicator spring stiffness constant, gives the average, or mean pressure, of the gases in the cylinder throughout the working stroke. This value is called the 'mean effective pressure' (symbol p_m).

Indicated horsepower

The piston load is given by:

$$\text{Mean effective pressure} \times \text{Piston area} = p_m \times A$$

The work per stroke is derived as the product of this load and the length of the stroke, in feet.

Then, if n' is the number of working strokes per minute

$$\text{Indicated horsepower} = \frac{p_m A l n'}{33{,}000}$$

or, Indicated power $= p_m A l n'_s$ watts (SI system)

Brake horsepower

This is measured by a rope or Prony brake, as explained in Chapter 7.

Mechanical efficiency

This is the ratio $\dfrac{\text{Brake horsepower}}{\text{Indicated horsepower}} = \eta_m$

or, $\dfrac{\text{Brake power (watts)}}{\text{Indicated power (watts)}} = \eta_m$

But, since brake horsepower is indicated horsepower minus frictional horsepower (fhp), then

$$\eta_m = 1 - \frac{\text{fhp}}{\text{ihp}}$$

or,
$$\eta_m = 1 - \frac{\text{Friction power (watts)}}{\text{Indicated power (watts)}}$$

Thermal efficiency

This is given by the expression $\frac{\text{Work output}}{\text{Heat input}}$

Work is converted into heat units by using the constants

1 Btu = 778 ft lbf of work
1 Chu = 1400 ft lbf of work
1 kilogramme calorie = 4186 joules
(1 gramme calorie = 4·186 joules)

Exercises

1. During a simple trial of a four stroke diesel engine, fitted with a rope brake, the following observations were made:

Cylinder bore	6 in.
Stroke	10 in.
Effective diameter of brake	31 in.
Brake load, including carrier	125 lbf
Spring balance reading	5 lbf
Speed	275 rev/min
Average m.e.p. from indicator card	110 lbf/in^2

Calculate (*a*) ihp (*b*) bhp (*c*) mechanical efficiency.

2. An oil engine uses 1·6 kg of oil per hour. The ip of the engine is 4·5 kW and the bp is 3·375 kW. If the calorific value of the oil used is 41,830 kJ/kg calculate the thermal efficiency based on (*a*) ip (*b*) bp.

3. An engine uses 0·40 lbf of oil per ihp hour. If the calorific value of the oil is 18,500 Btu/lbf calculate its indicated thermal efficiency.

4. A diesel engine under test gave a fuel consumption of 0·274 kg/kW. The fuel used had a calorific value of 42,300 kJ/kg. If the mechanical efficiency of the engine was 0·82, calculate (*a*) the brake thermal efficiency, (*b*) the indicated thermal efficiency.

5. A four stroke oil engine has a cylinder diameter of 7½ in. and a stroke of 14 in. During a test the following readings were taken:

Effective brake diameter	48 in.
Brake load	73 lbf
Spring balance reading	6 lbf
Average speed	325 rev/min

M.e.p. from indicator card	80 lbf/in^2
Oil consumption	0·09 lbf/min
Calorific value of fuel oil	18,500 Btu/lbf

Calculate (*a*) bhp (*b*) ihp (*c*) mechanical efficiency (*d*) brake thermal efficiency (*e*) indicated thermal efficiency.

6. The lubricating oil from an engine is cooled by passing it through a nest of tubes around which water is circulated. The oil flows at a rate of 17 litre/min and is cooled from 76°C to 28°C.

(*a*) Estimate the rate of flow of water required if it enters at 12°C and leaves at 28°C. 1 litre of oil is 0·9 kg and its specific heat is 1·874 kJ/kg °K.

(*b*) The heat carried away by the cooling water is 25 per cent of the total heat energy in the fuel used by the engine. Of this total energy 28 per cent is converted into useful work. Calculate the power developed by the engine. to burn 1 kg of the coal completely.

17

Combustion of fuels

Most substances are 'compounds' because they can be built-up from other substances by chemical processes. Some substances cannot be manufactured by making other substances combine, and these are known as 'elements'. Examples of elements which are solid at room temperature are aluminium, copper, iron, carbon and sulphur: examples of elements which are liquid at room temperature are bromine, iodine and mercury: examples of elements which are gaseous at normal temperatures and pressures are hydrogen, nitrogen and oxygen.

From the hundred or so elements which exist all known compounds are built up. However, the obvious properties of a compound bear no relation to the properties of the elements from which it is made. Take, for example, common salt. It is a compound formed from the elements sodium and chlorine. Sodium is a soft substance which can be cut with a knife, leaving a metallic surface which tarnishes at once on exposure to the air. If thrown into water it floats and decomposes the water so vigorously that combustion and even explosions may ensue. Chlorine is a poisonous gas, yet it combines readily with sodium to form common salt, which is used extensively with food, and dissolves in water quietly and without fuss.

Chemical compounds are formed by the combination of atoms of different elements. A few atoms—sometimes only one—of one element combine with a few atoms of another element to form a molecule of the compound. A molecule is the smallest quantity of the compound we can have without losing or changing any of its properties. For a few substances the atom and the molecule may be the same thing. That is, the molecule may consist of only one atom. On the other hand a molecule of oxygen, or hydrogen, consists of two atoms which are exactly alike but joined together. In other words we do not normally find oxygen or hydrogen atoms singly.

Chemical symbols

All elements are represented chemically by letters of the alphabet. Here are just a few:

Element	Carbon	Hydrogen	Nitrogen	Oxygen
Symbol	C	H	N	O

A number used as a suffix to the symbol indicates the number of atoms of each particular element in the molecule. Thus, a molecule of hydrogen contains two atoms, and is expressed chemically as H_2.

A compound is represented chemically by a combination of letters with numbers used as suffixes. Thus a molecule of carbon dioxide contains one atom of carbon and two atoms of oxygen combined together, and is expressed chemically as CO_2.

Fuels

A fuel is a substance which, once it has been raised to its ignition temperature, continues to burn without any external support, provided a sufficient quantity of oxygen is available.

Fuels may be solid, liquid or gaseous, and the principal constituents of any fuel are carbon and hydrogen.

Chemical equations

The combustion of a fuel may be represented by a chemical equation, which is a means of writing down, in terms of symbols and numbers, a chemical change which takes place when molecules of two or more substances combine together.

If we want to deal with a number of the same molecules we merely place the number in front of the chemical expression for the element or compound. Thus $2O_2$ represents two molecules of oxygen, each molecule containing two atoms; and $4CO_2$ represents four molecules of carbon dioxide, each molecule containing one atom of carbon and two atoms of oxygen.

We can then build up basic chemical processes or reactions as they are called, such as the following:

1. Carbon combines with oxygen and burns to form carbon dioxide if sufficient oxygen is available for complete combustion. Chemically this reaction is expressed:

$$C + O_2 = CO_2 \qquad 17.1$$

1 molecule of carbon + 1 molecule of oxygen
= 1 molecule of carbon dioxide

2. If insufficient oxygen is available, the combustion is not complete

and carbon monoxide is formed instead of carbon dioxide. This reaction is expressed:

$$2C + O_2 = 2CO \qquad 17.2$$

2 molecules of carbon + 1 molecule of oxygen
= 2 molecules of carbon monoxide

3. Hydrogen combines with oxygen to form steam, which is water in the form of a vapour. Thus,

$$2H_2 + O_2 = 2H_2O \qquad 17.3$$

2 molecules of hydrogen + 1 molecule of oxygen
= 2 molecules of steam or water

Note, in each of the above equations, and in all chemical equations, that the number of atoms on one side balance the number of atoms of the same element on the other side.

In every equation concerned with the combustion of a fuel, substantial quantities of oxygen are needed. Consequently, if the maximum amount of heat is to be obtained from the fuel, an adequate supply of oxygen is essential. This oxygen is obtained from air in the atmosphere, although a very small amount may exist in the fuel.

Atmospheric air

Atmospheric air consists primarily of oxygen and nitrogen. We can neglect all other substances when dealing with combustion problems.

The proportions of oxygen and nitrogen in the air may be considered either by weight or by volume.

By weight, air contains 23 per cent oxygen and 77 per cent nitrogen.

By volume, air contains 21 per cent oxygen and 79 per cent nitrogen.

Only oxygen is used in the combustion of a fuel; nitrogen taking no part in the chemical changes which occur.

Atomic weight

It is convenient to regard a volume of 179·5 cubic feet as the standard volume for a gas measured at a temperature of 32°F and a pressure of 14·7 lbf/in^2 absolute. The weight of a gas which occupies this volume under the standard conditions of temperature and pressure is known as its "atomic weight". Thus 1 lbf of hydrogen gas occupies this volume at standard temperature and pressure (abbreviated to S.T.P.) and so its atomic weight is said to be 1. Similarly 16 lbf of oxygen gas occupy the same volume at S.T.P., and so the atomic weight of oxygen is said to be 16. If measurements are taken in SI units, then 1 kgf of hydrogen occupies 11·2 m^3 at S.T.P.

The atomic weights of all the elements are relative, and obtained by comparison. At one time the hydrogen atom, which is the lightest of them all, was regarded as the standard for comparison. Later the hydrogen atom was replaced by the oxygen atom as the standard, but the atomic weights of the elements remained the same. More recently, the carbon atom, with an atomic weight of 12, has become the standard for comparison of all other atomic weights. We do not, however, need to worry about these changes.

All atomic weights are rounded off to the nearest whole number. We need only concern ourselves with the following:

Element	Carbon	Hydrogen	Oxygen	Nitrogen
Symbol	C	H	O	N
Atomic weight	12	1	16	14

Molecular weights

We have seen that a molecule of an element such as hydrogen, oxygen and nitrogen contains two atoms, and so the molecular weights of these elements are twice their respective atomic weights.

The molecular weight of a compound is obtained by adding up the weights of each of the atoms of the elements of which it is composed. Thus, carbon dioxide, chemically expressed as CO_2, contains one atom of carbon and two of oxygen. Because the atomic weight of carbon is 12, and the atomic weight of oxygen is 16, the molecular weight of carbon dioxide is $12 + 2(16) = 12 + 32 = 44$.

The combustion of carbon and hydrogen

In all combustion problems oxygen is needed. This oxygen is obtained from the atmosphere and it combines with carbon and with hydrogen in what are known as chemical reactions. While these reactions are taking place heat energy is given out.

(*a*) *Complete combustion of carbon*

We have seen (equation 17.1) that the chemical equation for this reaction is:

$$C + O_2 = CO_2$$

Applying the atomic weights given in the above table:

$$12 + 2(16) = 12 + 32 = 44$$

That is,

12 lb of carbon combine with 32 lb oxygen to produce 44 lb of carbon DIoxide, and so, by simple proportion, 1 lb of carbon combines with $\frac{32}{12}$ lb oxygen to produce $\frac{44}{21}$ lb of carbon DIoxide.

(b) Incomplete combustion of carbon
In practice the combustion of carbon is very rarely complete.

If there is insufficient oxygen, the reaction of carbon with oxygen is expressed by the equation:

$$2C + O_2 = 2CO$$

The molecular weights can again be obtained from the table of atomic weights, thus:

$$\begin{aligned}(2 \times 12) + (2 \times 16) &= 2(12 + 16)\\ 24 \quad + \quad 32 \quad &= \quad 56\end{aligned}$$

That is,

24 lb of carbon combine with 32 lb of oxygen to produce 56 lb of carbon MONoxide, or
1 lb of carbon combines with $\frac{32}{24}$ lb of oxygen to produce $\frac{56}{24}$ lb of carbon MONoxide.

(c) Combustion of hydrogen
This reaction is expressed as

$$2H_2 + O_2 = 2H_2O$$

The molecular weights are

$$\begin{aligned}2(2) + 2(16) &= 2(2 + 16)\\ 4 \quad + \quad 32 \quad &= \quad 36\end{aligned}$$

That is,

4 lb of hydrogen combines with 32 lb of oxygen to produce 36 lb of steam (or water),
$\therefore$ 1 lb of hydrogen combines with $\frac{32}{4}$ lb of oxygen to produce $\frac{36}{4}$ lb of steam.

Minimum weight of air required

If the weights of the combustible elements—carbon and hydrogen-are known, then the theoretical weights of oxygen required to burn them can be obtained by simple proportion as shown above. The total weight of oxygen is obtained by adding the separate weights of oxygen required in each combustion equation.

Then, because air contains only 23 per cent oxygen by weight, the weight of air required is

$$\frac{100}{23} \text{ times this weight of oxygen}$$

This amount of air is known as the minimum, or theoretical, weight of air for combustion.

Excess air

In practice, more than the theoretical amount of air is necessary in order to secure complete combustion. This is because all the air does not come into intimate contact with the particles of fuel. The amount supplied varies with the type of fuel and the firing conditions, but may exceed the theoretical amount by 25 or even 50 per cent.

Products of combustion

The greater part of the air used to supply the oxygen needed for combustion consists of nitrogen gas. This gas takes no part in the chemical reactions which take place, and so all the nitrogen present in the air supply passes through the combustion chamber and emerges with the exhaust gases.

The composition of the exhaust products is therefore, as follows:

1. Carbon dioxide, if the combustion is complete.
2. A mixture of carbon dioxide and carbon monoxide, if the combustion of carbon is incomplete.
3. Steam, from the combustion of hydrogen.
4. Nitrogen, from the air supply.
5. Oxygen, if an excess of air—that is, more than the minimum amount required—is supplied.

The weight of each of these products can be determined from the weights of combustible matter and air supplied, and the ensuing chemical reactions. Each is usually then changed into the form of a percentage of the total weight of exhaust gases.

When solid fuels are burnt, a small percentage of the fuel is usually found to be incombustible. This remains as ash or clinker in the grate, and does not pass up the chimney with the gaseous products of combustion.

Example

The analysis, by weight, of a fuel supplied to a boiler was carbon 85 per cent, hydrogen 8 per cent and the remainder incombustible matter. Determine (*a*) the theoretical weight of air required for complete combustion of 1 kg of the fuel, and (*b*) the percentage composition, by weight, of the flue gases if the actual air supplied was 15 kg per 1 kg of fuel.

Solution

(*a*) For complete combustion, carbon combines with oxygen to form carbon dioxide, thus

$$C + O_2 = CO_2$$

The molecular weights are

$$12 + 2(16) = 44$$

That is, 12 kg carbon combine with 32 kg oxygen to produce 44 kg carbon dioxide. But, in 1 kg of fuel there is 0·85 kg carbon. By proportion, 0·85 kg carbon combines with $\frac{32}{12} \times 0{\cdot}85$ kg oxygen $= 2{\cdot}27$ kg oxygen to produce $\frac{44}{12} \times 0{\cdot}85$ kg carbon dioxide $= 3{\cdot}12$ kg carbon dioxide.

Hydrogen combines with oxygen to form steam, thus

$$2H_2 + O_2 = 2H_2O$$

The molecular weights are

$$2(2) + 2(16) = 36$$

That is, 4 kg hydrogen combine with 32 kg oxygen to produce 36 kg steam. But, in 1 kg of fuel there is 0·08 kg hydrogen. By proportion, 0·08 kg hydrogen combines with $\frac{32}{4} \times 0{\cdot}08$ kg oxygen $= 0{\cdot}64$ kg oxygen to produce $\frac{36}{4} \times 0{\cdot}08$ kg steam $= 0{\cdot}72$ kg steam.

Total amount of oxygen required for the combustion of carbon and hydrogen in 1 kg fuel is

$$2{\cdot}27 + 0{\cdot}64 = 2{\cdot}91 \text{ kg}$$

This is the amount of oxygen which is taken from the air to burn each 1 kg of fuel.

But 100 kg air contains 23 kg oxygen.

$\therefore$ Theoretical weight of air required for the complete combustion of 1 kg of fuel $= \frac{100}{23} \times 2{\cdot}91 = 12{\cdot}65$ kg

(*b*) Actual air supplied for each 1 kg of

$$\text{Fuel burnt} = 15 \text{ kg}$$

This contains $\frac{77}{100} \times 15 = 11{\cdot}55$ kg nitrogen, all of which passes out of the chimney with the flue gases.

$$\text{Excess air supplied} = 15 - 12{\cdot}65 = 2{\cdot}35 \text{ kg}$$

This contains $\frac{23}{100} \times 2{\cdot}35 = 0{\cdot}54$ kg excess oxygen per 1 kg of fuel.

The flue gases will contain:

1. CO_2 formed by burning the carbon,
2. N_2, which is an incombustible gas, from ALL the air supplied,
3. H_2O as steam, formed by burning the hydrogen,
4. excess O_2.

Constituent	*Mass (kg)*	*Per cent*
CO_2	3·12	19·55
N_2	11·55	72·55
H_2O	0·72	4·51
Excess O_2	0·54	3·39
	15·93	100·0

Example

A fuel has the following analysis by weight: carbon 81 per cent; hydrogen 8 per cent; oxygen 7 per cent; the remainder being ash. Find, from first principles, the weight of air required for the complete combustion of 1 lb of this fuel. If the total products of combustion per 1 lb of this fuel weigh 16·7 lb, find the percentage excess air supplied.

(U.L.C.I.)

(Atomic weights C 12; O 16; H 1. Air contains 23 per cent by weight of oxygen.)

Solution

Carbon	Hydrogen
$C + O_2 = CO_2$	$2H_2 + O_2 = 2H_2O$
$12 + 32 = 44$	$4 + 32 = 36$
$3 + 8 = 11$	$1 + 8 = 9$

0·81 lb of carbon requires $(0{\cdot}81 \times \frac{8}{3}) = 2{\cdot}16$ lb oxygen

0·08 lb hydrogen requires $(0{\cdot}08 \times 8) = 0{\cdot}64$ lb oxygen

Total requirement is

$$2{\cdot}16 + 0{\cdot}64 = 2{\cdot}80 \text{ lb oxygen}$$

$$\text{Oxygen required from air} = 2{\cdot}80 - 0{\cdot}07 = 2{\cdot}73 \text{ lb}$$

$$\text{Air required for complete combustion} = 2{\cdot}73 \times \tfrac{100}{23}$$

$$= 11{\cdot}87 \text{ lb}$$

$$\text{Total essential products of combustion} = 11{\cdot}87 + 1 = 12{\cdot}87 \text{ lb}$$

$$\text{Excess air} = 16{\cdot}7 - 12{\cdot}87 = 3{\cdot}83 \text{ lb}$$

$$\text{Percentage excess air} = \frac{3{\cdot}83}{11{\cdot}87} \times 100$$

$$= 32{\cdot}2 \text{ per cent}$$

Summary

The burning of a fuel is the process of chemical combination of its elements with the oxygen of the air.

It is important to know that in chemical reactions the smallest particles of the elements which take part in the reactions are called 'molecules'. These may consist of one or more atoms.

Chemical symbols and atomic weights

We shall be concerned at this stage of the course with the elements carbon (symbol C), hydrogen (H), nitrogen (N), and oxygen (O). Their atomic weights are carbon 12, hydrogen 1, nitrogen 14, and oxygen 16.

The combustion equations. Combining weights

The combustion equations give the atomic quantities in the reacting molecules. Proportional weights of the elements and compounds in the reaction can immediately be deduced, as below.

Complete combustion of carbon:

$$C + O_2 = CO_2$$
$$12 + 32 = 44$$

Incomplete combustion of carbon:

$$2C + O_2 = 2CO$$
$$24 + 32 = 56$$

Combustion of hydrogen:

$$2H_2 + O_2 = 2H_2O$$
$$4 + 32 = 36$$

From these equations the weight of oxygen necessary to burn a given weight of the fuel is easily found.

Weight of air required

Air contains 23 per cent by weight of oxygen, the remainder being nitrogen. To find the air needed to burn a fuel, then, we write

$$\text{Air required, by weight} = \text{Oxygen required} \times \tfrac{100}{23}$$

Excess air

In practice some excess air must be admitted to allow combustion to be as complete as circumstances permit. Questions often ask for the proportion of this excess air. This is given by the expression:

$$\frac{\text{Air admitted} - \text{Necessary air}}{\text{Necessary air}} \times 100 \text{ per cent}$$

The products of combustion. Ash

It should be remembered that not all of the reacting elements become gases or vapours and pass up the chimney. Some 'ash' is left by solid fuels. It must often be allowed for in calculations.

Exercises

1. A coal contains 87 per cent of carbon and 4 per cent of hydrogen, the remainder being incombustible. Find the minimum weight of air necessary to burn 1 kg of the coal completely.

2. A coal contains 85 per cent carbon and 5 per cent hydrogen, the remainder being incombustible. If the minimum amount of air for complete combustion is supplied, find the percentage composition of the flue gases (*a*) if the moisture is extracted before the gases reach the chimney (*b*) if no provision is made for drying.

3. The analysis, by weight, of a fuel supplied to a boiler was carbon 83 per cent, hydrogen 8 per cent, oxygen 2 per cent, and the remainder incombustible matter. Determine: (*a*) the theoretical weight of air required for complete combustion of 1 lbf of the fuel (*b*) the percentage composition, by weight, of the dry flue gases if the actual air supplied was 15 lbf per 1 lbf of fuel.

4. A sample of oil contains 83 per cent carbon and 17 per cent hydrogen. Determine the theoretical weight of air for complete combustion of 1 kg of oil.

5. The analysis of a boiler fuel showed that it contained 87 per cent carbon, 3 per cent hydrogen and 4 per cent oxygen. Determine, from first principles, the weight of air theoretically required for complete combustion of 1 lbf of coal.

6. The percentage analysis, by weight, of a certain fuel is carbon 80, hydrogen 8, oxygen 8, and the remainder incombustible matter. If 50 per cent excess air is used for complete combustion of the fuel, determine the percentage analysis of the flue gas.

7. The analysis, by weight, of a dry boiler fuel was carbon 82 per cent, hydrogen 8 per cent, and the remainder incombustible matter. Assuming complete combustion of the fuel, calculate

(*a*) the minimum weight of air required per 1 kg of fuel supplied.

(*b*) the percentage composition of the dry flue gases, by weight, if the actual air supplied was 16 kg per 1 kg of fuel.

8. A heavy fuel oil containing 85 per cent carbon, 10 per cent hydrogen, 2 per cent oxygen and 3 per cent impurities, by weight, is burnt in an engine. Calculate the percentage composition of the exhaust gases if 50 per cent excess air is used for combustion.

9. A sample of fuel gave the following analysis, by weight: carbon 91·7 per cent; hydrogen 4 per cent; oxygen 2·9 per cent; remainder ash. Find

(*a*) the minimum amount of air required for the complete combustion of 1 lbf of fuel.

(*b*) the amount of carbon dioxide expressed as a percentage by weight of the dry flue gas. (Atomic weights: C = 12, O = 16, H = 1. Air contains 23 per cent by weight of oxygen.) (U.L.C.I.)

18

Liquids at rest

All substances can be classified under one of three headings to denote their state, namely, solid, liquid or gaseous.

Solids are substances which retain their shape and size under ordinary circumstances, and unless acted upon by some large forces.

Liquids do not have any shape of their own, but assume the shape of the vessel into which they are poured. They do, however, have a definite volume.

Gases have no definite shape of their own, and also have no definite volume. In fact, if they are placed in a vessel which can be sealed, they will occupy the whole of the internal volume of the vessel.

In this chapter we propose dealing with the properties of liquids which are at rest, and their behaviour when subjected to certain forces. Although we shall define properties and state laws which can be applied to ALL liquids, we shall for the most part be thinking of only one liquid, namely, water.

We shall, first of all, define certain properties. It is not necessary at this stage to give precise definitions as this would involve a more extensive knowledge of physics than we need here. Our definitions, however, will be sufficient to enable us to understand the problems we are about to face.

Density

The density of a liquid may be expressed in two ways—either as a specific weight or as a specific mass.

We shall choose the former and state that

$$\text{Density} = \text{Specific weight} = \frac{\text{Weight}}{\text{Volume}}$$

The Imperial system measures weight in lbf and volume in ft^3, so the units of density will be lbf/ft^3 and our symbol will be w.

Alternatively, we can obtain a mass density by determining the mass in kilogrammes and the volume in cubic metres, giving the mass density in kg/m^3 units.

Relative density

It is useful to compare the densities of two liquids; or, of course, the weights of two liquids which occupy equal volumes.

If we select distilled water at a temperature of 4°C as our standard for comparison a ratio is obtained which is called 'relative density'. Thus,

$$\text{Relative density} = \frac{\text{Specific weight of liquid}}{\text{Specific weight of water}}$$

Viscosity

One of the essential properties of a liquid is that it shall flow. Water, being a 'thin' liquid will flow much more easily than a 'thick' liquid such as heavy oil. In other words, the heavy oil resists any forces which tend to change its shape or position much more than does water. This property of a liquid is called its viscosity.

Incompressibility

We have already seen that a liquid occupies a definite volume under stated conditions. In other words, a liquid resists very strongly any attempts by external forces to change that volume. Thus, if an external compressive force is exerted on a liquid, the change in its volume is so minute that it can be neglected. Consequently, we shall say that a liquid is incompressible.

Surface tension

A large mass of liquid does not keep its shape unless it is held in a rigid container. Very small amounts of liquid, however, assume a definite form, and retain it at least temporarily. Thus drops of oil in water, bubbles of air in water, and drops of water falling freely from a dripping tap assume definite shapes, as if they were enclosed in skins. The skin, which forms a curved surface of contact between two liquids of different density, or between a liquid and a gas, is called a 'meniscus'. The meniscus is under the influence of a tensile force, the surface tension being the force measured in a direction tangential to the curved surface.

Suppose that we have a liquid in a vertical tube, the upper end of which is open to the atmosphere. The surface of the liquid in the tube, being in contact with the atmosphere, which has a different density, will be curved to form a meniscus. The shape of the meniscus is usually concave—in other words, the surface sags at the middle. If the tube contains mercury, however, the meniscus is convex. That is, the usual curvature is reversed.

Pressure

Pressure has already been defined as the force per unit area. That is,

$$p = \frac{F}{A}$$

If the force is determined in pounds and the area in square feet, the units of p are lbf/ft^2.

If the force is determined in newtons and the area in square metres, the units of p are then N/m^2.

Pressure at a point

If a liquid is at rest, the pressure at any point in the liquid is the same in all directions. This is shown diagrammatically in figure 18.1.

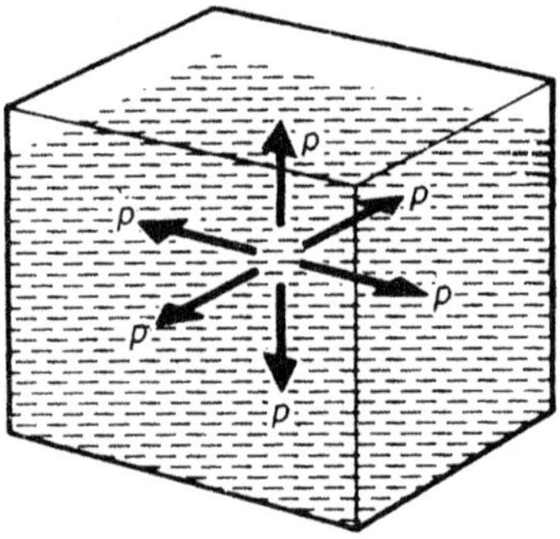

Figure 18.1 Pressure at a point in a liquid.

Pressure due to the weight of a liquid

Consider a tank of rectangular section, as in figure 18.2, and filled with water, the surface being exposed to the atmosphere. Such a surface will be level and is called a "free" surface.

Let its area be A ft^2.

If w be the weight of unit mass of water, then the force on the base of the tank due to the weight of the water above it is given by

$$P = wAh \qquad 18.1$$

where h = Depth of the water.

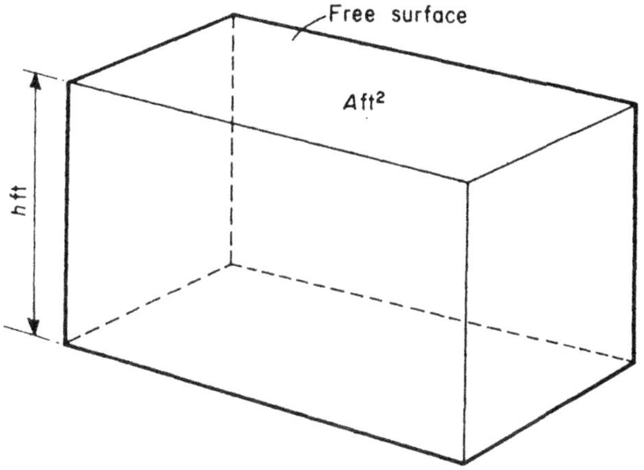

Figure 18.2 Pressure due to the weight of a liquid.

The pressure at any point on the base is

$$p = \frac{P}{A}$$

or, $$p = wh \qquad 18.2$$

If, of course, the mass is determined in kg, and the area in m², the pressure will be obtained in kgf/m². If pressure is required in N/m², the pressure in kgf/m² must be converted by multiplying by 9·81 N/kgf.

Similarly, at any other level, it can be shown that

$$p_1 = wh_1$$

where h_1 = Depth of water above the horizontal plane under consideration.

Because the pressure at a point is dependent on the weight of water above it, its value is constant on any horizontal plane.

When a liquid is at rest, the pressure at a point is the same in all directions. Consequently, the PRESSURE on all four sides of the tank must be constant at a given depth.

Do not be confused here with force, which equals pressure times area.

Total pressure

The pressure we have just determined is entirely due to the weight of water. At the free surface the pressure due to the weight of water above it is zero, but a pressure exists due to the effect of the Earth's atmosphere.

Consequently, we shall say, at the free surface,

$$\text{Total pressure} = \text{Atmosphere pressure}$$
$$= p_{at}$$

At a depth h feet below the free surface,

$$\text{Total pressure} = p_{at} + wh$$

In other words, the pressure determined from the expression $p = wh$ is a gauge pressure.

We shall not often be concerned with a total pressure.

Pressure on the sides of a tank

We have just seen that, at any point, due to the weight of water above it, the pressure is the same in all directions.

Consequently, at a point on the sides of a tank containing water at rest, the direction of the resultant pressure is normal to the surface. If it were not so, there would be movement of the water. This means that, if we cut a small hole in the side of a tank, water would gush out in a direction perpendicular to the surface at that point.

Thus, in figure 18.3, at each hole the direction of the jet of water

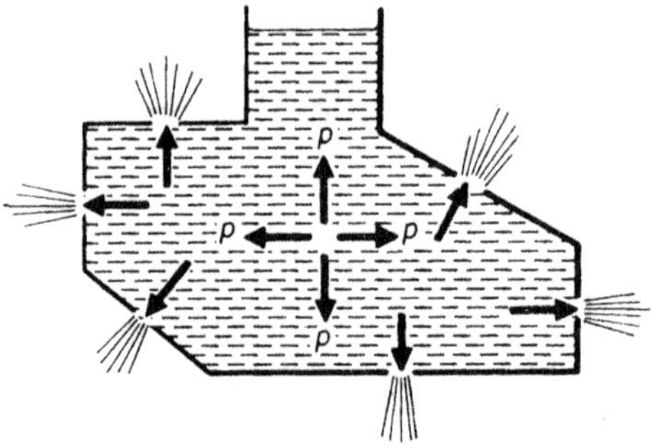

Figure 18.3 Pressure on the sides of a tank.

emerging from the tank would, initially be perpendicular to the surface of the tank.

Total thrust

Let ABCD be a rectangular lamina suspended vertically in a tank of water so that it is completely submerged, but the edge AB is in the free surface of the water as shown in figure 18.4.

Let A be the surface area of the lamina.

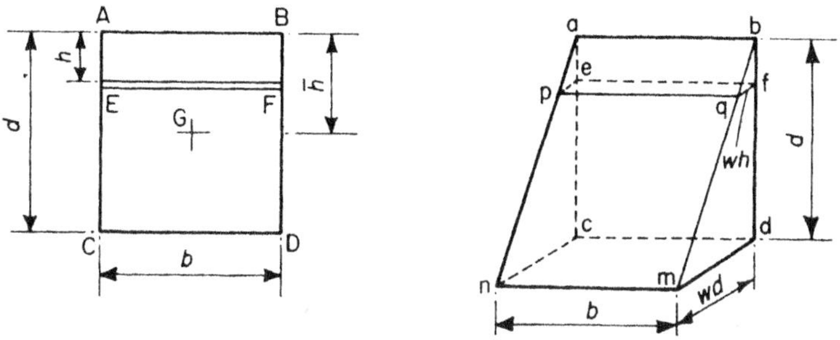

Figure 18.4 The pressure—depth diagram.

At a depth h the pressure at any point in the horizontal plane through EF is expressed by

$$p = wh$$

Consequently, a pressure-depth diagram can be drawn, such as that shown on the right hand side of figure 18.4, where the pressure ($p = wh$) is drawn perpendicular to the lamina at a depth h.

Because the pressure is directly proportional to the depth, the diagram takes the form of a prism.

The pressure at a point is the same in ALL directions, so that there will be a thrust on the vertical lamina, perpendicular to its surface.

At the surface,

Pressure due to the weight of water = 0

At a depth h,

pressure due to the weight of water = wh

That is,

$$\text{Average pressure between AB and EF} = \frac{wh}{2}$$

Force on the lamina between AB and EF

= Area ABEF × Average pressure on this area
= Volume of pressure-depth diagram above the horizontal plane through EF
= Volume abfepq in figure 18.4

At a depth d,

Pressure due to the weight of water = wd

$$\text{Average pressure between AB and CD} = \frac{wd}{2}$$

Total force, or thrust, on lamina ABCD

$$= \text{Area ABCD} \times \text{Average pressure}$$

$$= A \times \frac{wd}{2}$$

$$= wA \times \frac{d}{2}$$

But $\frac{d}{2}$ is the depth of the centroid of the rectangular lamina below the free surface, or,

$$\frac{d}{2} = \bar{h}$$

$\therefore$ Total force, or thrust, on lamina ABCD $= wA\bar{h}$ 18.3

This expression may be re-written

Total force, or thrust $= w\bar{h} \times A$

= Pressure at the centroid $\times$ Wetted area

= Volume abdcnm

(*Note:* The above expression only gives the magnitude of the total force on the lamina, and NOT the point where it would act if it were regarded as a single force. It does NOT act at the centroid of the lamina.)

Centre of Pressure of a submerged rectangular lamina

In figure 18.4 it was shown that the total force, or thrust, on the face of a submerged lamina could be expressed in terms of the volume of a prism erected on the lamina as a base.

The distance of the centre of gravity of this prism from the free surface of the liquid is known as the "centre of pressure". Because the prism is made up of a large number of vertical triangular laminae with their apices in the free surface, and the distance of the centroid of a triangular lamina is two-thirds of its vertical height from the apex, we can say that the distance of the centre of gravity of the prism is $\frac{2}{3}d$ vertically below the free surface of the liquid. From symmetry it must also be mid-way between the two triangular faces.

That is, the line of action of the centre of pressure passes through a point $\frac{2}{3}d$ below the free surface and midway between the two vertical edges of the lamina.

It has already been stated that the centre of pressure is the same as the centre of gravity of the prism which represents the forces acting on the submerged lamina.

Hence, the line of action of the total force, or thrust, on the lamina passes through the centre of pressure, or the centre of gravity of the pressure-depth diagram.

Transmission of fluid pressure in a pipe

Suppose we have an apparatus such as that shown by a line diagram in figure 18.5. In this apparatus two cylinders of different diameters are joined by a pipe filled with a liquid.

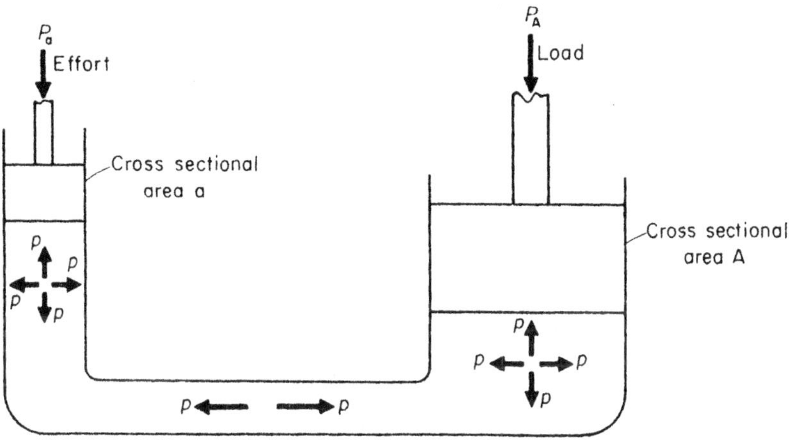

Figure 18.5 Transmission of fluid pressure.

Suppose that the two cylinders are sealed by pistons of cross-sectional areas a and A, where A is much larger than a. The shape and size of the pipe joining the two cylinders does not concern us here.

Suppose an external force P_a is applied to the small piston. As a result, the pressure exerted by the piston on the liquid is expressed by

$$p = \frac{P_a}{a}$$

If we assume the liquid to be incompressible, this pressure must be transmitted throughout the liquid.

If the apparatus is shallow we can also assume that the weight of the liquid in the cylinders does not appreciably affect the fluid pressure. Consequently, we can say that the pressure p exerted by the force P_a on the small piston is constant throughout the apparatus and will therefore be the pressure of the liquid exerted on the bottom of the large piston of cross-sectional area A. Hence,

$$\begin{aligned} \text{Force on large piston} &= p \times A \\ &= \text{Load which can be raised by piston } A \\ &= P_A \end{aligned}$$

That is, $$pA = P_A$$

or, $$p = \frac{P_A}{A}$$

But $$p = \frac{P_a}{a}$$

$$\therefore \quad \frac{P_A}{A} = \frac{P_a}{a}$$

or, transposing, we get

$$\frac{P_A}{P_a} = \frac{A}{a}$$

But $$\frac{P_A}{P_a} = \frac{\text{Load being raised by piston } A}{\text{Effort being applied to piston } a}$$

$$= \text{Mechanical advantage of apparatus}$$

But $$\frac{P_A}{P_a} = \frac{A}{a}$$

$$\therefore \quad \text{Mechanical advantage} = \frac{A}{a}$$

That is, by a suitable choice of areas (and hence diameters), we can obtain a substantial mechanical advantage by the use of fluid pressure. This principle is used in a number of hydraulic machines, including presses, and jacks.

Example

A lock gate is 15 ft wide and has water on both sides. The depth of water on one side is 12 ft and on the other side is 4 ft.

(*a*) Calculate the thrust on each side of the gate and the magnitude of the resultant thrust. Assume water weighs 62·5 lbf/ft^3.

(*b*) Determine also the height of the line of action of the resultant thrust above the bottom of the lock gate.

Solution

$$\text{Thrust} = \text{Specific weight} \times \text{Wetted area} \times \text{Depth of centroid of wetted area}$$

$$= wA\bar{h}$$

$$= wA\frac{d}{2}$$

(*a*)

$$\text{Thrust on deep side} = P_1 = 62{\cdot}5 \times (15 \times 12) \times \tfrac{12}{2}$$
$$= 67{,}500 \text{ lbf}$$
$$= 30{\cdot}1 \text{ tonf}$$

$$\text{Thrust on shallow side} = P_2 = 62{\cdot}5 \times (15 \times 4) \times \tfrac{4}{2}$$
$$= 7500 \text{ lbf}$$
$$= 3{\cdot}35 \text{ tonf}$$

Resultant thrust on lock gate $= P_1 - P_2$ from left to right
$= 30{\cdot}1 - 3{\cdot}35$
$= 26{\cdot}75$ tonf

(*b*)

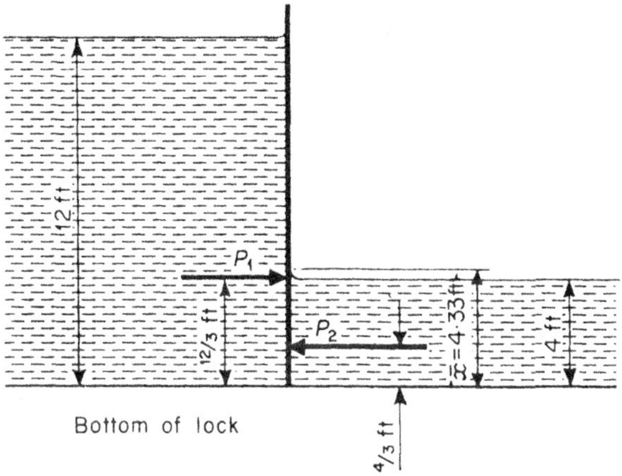

Figure 18.6 A lock gate.

Thrust on deep side acts at a height of $\frac{12}{3}$ ft from the bottom of the lock.

Thrust on shallow side acts at a height of $\frac{4}{3}$ ft from the bottom of the lock.

Suppose the line of action of the resultant thrust $(P_1 - P_2)$ is at a height of $\bar{x}$ ft above the bottom of the lock.

Take moments about the bottom of the lock. Then,

$$(67{,}500 \times \tfrac{12}{3}) - (7500 \times \tfrac{4}{3}) = (60{,}000 \times \bar{x})$$

or,

$$(30{\cdot}1 \times 4) - (3{\cdot}35 \times \tfrac{4}{3}) = (26{\cdot}75 \times \bar{x})$$

$$120{\cdot}4 - 4{\cdot}47 = 26{\cdot}75\bar{x}$$

$$\bar{x} = \frac{115{\cdot}93}{26{\cdot}75}$$

$$= 4{\cdot}33 \text{ ft}$$

Example

In a hydraulic machine a force of 400 N is applied to a piston having a diameter of 25 mm. Pressure is transmitted through a fluid to a larger piston having a diameter of 300 mm. Calculate (*a*) the magnitude of the pressure transmitted, and (*b*) the load which the larger piston can carry assuming the machine has an efficiency of 100 per cent.

Solution

Let P_a be the effort applied to the smaller piston of area a, and P_A be the load raised by the larger piston of area A.

(*a*) Then, if p is the pressure given to the fluid by piston a and transmitted throughout the fluid

$$p = \frac{P_a}{a}$$

$$= \frac{400}{\frac{\pi}{4} \times (0{\cdot}025)^2}$$

$$= 814{\cdot}871 \text{ kN/m}^2$$

(*b*) If the machine is perfect—that is, its efficiency is 100 per cent.

$$\frac{P_a}{a} = \frac{P_A}{A}$$

or,

$$P_A = P_a \times \frac{A}{a}$$

$$= 400 \times \frac{\frac{\pi}{4} \times 300^2}{\frac{\pi}{4} \times 25^2}$$

$$= 57{,}600 \text{ N}$$

$$= 57{\cdot}6 \text{ kN}$$

Example

In a small hydraulic machine the piston diameters are 2 and 20 centimetres respectively. A mass of 5 kg is loaded onto the smaller of the two pistons, and pressure is transmitted through a fluid to the larger piston. Calculate (*a*) the magnitude of the pressure transmitted in newtons per square metre, and (*b*) the maximum load which can be carried by the larger piston.

Assume the efficiency of the machine to be 100 per cent.

Solution

Let $\quad P_a$ = Effort applied to smaller piston of area a.

and $\quad P_A$ = Load raised by larger piston of area A.

(*a*) If p = Pressure given to fluid by piston a and transmitted throughout the fluid.

$$p = \frac{P_a}{a}$$

$$= \frac{5 \times 9{\cdot}81 \text{ N/m}^2}{\frac{\pi}{4} \times \left(\frac{2}{100}\right)^2}$$

But since 1 kg weighs 9·81 newtons

$$p = \frac{5 \times 9{\cdot}81 \times 10{,}000}{\pi} \text{ newtons/m}^2$$

$$= 156{,}000 \text{ newtons/m}^2$$

(*b*) If the efficiency of the machine is 100 per cent

$$\frac{P_a}{a} = \frac{P_A}{A}$$

or,

$$P_A = P_a \times \frac{A}{a}$$

$$= 5 \times \frac{\frac{\pi}{4} \times 20^2}{\frac{\pi}{4} \times 2^2}$$

$$= 5 \times 10^2$$

$$= 500 \text{ kg}$$

A note on Archimedes' principle

Most students will have studied the principle of Archimedes in earlier courses, but for revision purposes it is re-stated here.

'The apparent loss in weight of a body when weighed first in air and then in a liquid is equal to the weight of the liquid displaced.'

Buoyancy and flotation

When a body floats partially submerged, Archimedes' principle still applies. The submerged portion displaces liquid equal to its own volume, and it is the weight of this displaced liquid which is equal to the whole weight of the floating body.

Example

(*a*) State the principle of Archimedes.

(*b*) (i) A body is found to float in an oil with 0·6 of its volume submerged and in water with 0·5 of its volume immersed. Find the specific gravity of the oil.

(ii) An open ended U tube whose arms are equal length and bore uniform, is arranged vertically and filled with water until the level in the arms is 8 inches from the upper ends. A quantity of the oil, which does not mix with water, is introduced into one arm until it is completely filled. Find the height of the oil column. (N.C.T.E.C. 1966)

Solution

(*a*) As above.

(*b*) (i) Specific gravity of oil = Relative density

$$= \frac{\text{Volume of water displaced by solid}}{\text{Volume of liquid displaced by solid}}$$

$$= \frac{0{\cdot}5}{0{\cdot}6}$$

$$= 0{\cdot}833$$

(ii) If h is the height of the oil column, then $\frac{5}{6}h$ will be the height of the water column balancing it, and $\frac{1}{6}h$ the air remaining in the tube above the water. Then

$$h + \tfrac{1}{6}h = 2 \times 8 \text{ in.}$$

or,

$$\tfrac{7}{6}h = 16$$

$$\therefore\ h = \frac{6 \times 16}{7}$$

Height of oil column = 13·71 in.

Summary

The properties of liquids

These include density, relative density (or specific gravity), viscosity, incompressibility, and surface tension.

Pressure in a liquid

The pressure in a liquid is defined as the force per unit area.

$$p = \frac{P}{A}$$

The pressure at a point in a liquid is the same in all directions, because fluids transmit pressure equally in all directions.

Pressure due to the weight of a liquid

If w is the density of the liquid, and h the height of the liquid standing over the point, then

$$\text{Pressure at a point} = wh$$

Total pressure of a liquid

This is the sum of the atmospheric and liquid pressures.

$$\text{Total pressure} = p_{\text{at}} + wh$$

Total thrust

The pressure on the surfaces of containers of liquids is determined by summing the loads on elementary areas of the surface. These loads are all different, and depend on the liquid pressure at the element under consideration. Pressure is proportional to depth, so the pressure at mid-depth is the average pressure over the whole surface.

For regular surfaces of area A

$$\begin{aligned}\text{Total thrust} &= \text{Area} \times \text{Average pressure} \\ &= A \times \frac{wd}{2} \\ &= wA \times \frac{d}{2}\end{aligned}$$

Centre of pressure of a submerged rectangular lamina

If a lamina, submerged in a liquid, can be considered to be hinged at the liquid surface line, then the liquid pressure will tend to swing the lamina on this hinge.

The total force, or thrust, can be expressed in terms of the volume of a prism erected on the lamina as a base. The centre of pressure is then the same as the centre of gravity of the prism.

Transmission of fluid pressure in a pipe

Fluids transmit pressure equally in all directions. This principle is used in many hydraulic machines, particularly those concerned with lifting.

If a liquid pressure p acts on two pistons of areas A and a, the pistons being loaded with P_A and P_a respectively, then

$$p = \frac{P_A}{A} = \frac{P_a}{a}$$

The mechanical advantage of the machine is $\dfrac{A}{a}$.

Archimedes' principle. Buoyancy and flotation

The apparent loss in weight of a body when weighed first in air and then in a liquid is equal to the weight of the liquid displaced.

In considering problems of buoyancy and flotation of partially immersed bodies, note that the volume of the liquid displaced is the same as the immersed volume of the body.

The relative density, or specific gravity, of a liquid can be found by comparing the loss in weight of a body immersed in it with the loss in weight of the same body immersed in water, for these losses in weight correspond with the weights of equal volumes of the liquid and water displaced by the body.

Exercises

(Assume 1 m^3 of water = 10^3kg.)

1. A dock gate, 40 ft wide, has sea water on one side to a depth of 30 ft. Calculate the horizontal force, in tonf, on the face of the gate. Sea water weighs 64 lbf/ft^3.

2. The vertical sides of an oil tank are 3 m wide, and the depth of oil in the tank is 2·5 m. Calculate the horizontal thrust on one side of the tank, if the relative density of the oil is 0·8.

3. A water tank is 50 ft long, 25 ft wide, and 10 ft deep. If the depth of water in the tank is 7 ft, determine
   (*a*) the pressure on the base of the tank
   (*b*) the pressure at the centroid of the immersed portion of a side
   (*c*) the total force on the base
   (*d*) the total force on each end
   (*e*) the total force on each side
   Water weighs 62·5 lbf/ft^3.

4. A vertical lock gate is 12 ft wide and has water on both sides. The depth of water on one side is 12 ft and on the other 4 ft. Calculate (*a*) the thrust on each side of the gate (*b*) the resultant thrust, and the height at which it acts. Assume 1 ft^3 of water weighs 62·5 lbf.

5. A vertical lock gate is 5 m wide and has water on both sides. The depth of water on one side is 6 m and on the other side 4 m. Calculate the resultant horizontal thrust on the gate and the distance from its base at which this thrust acts.

6. A vertical lock gate 13 m wide has water on both sides. On one side the depth of the water is 10 m, and on the other side the depth is 2 m. Calculate the resultant horizontal force on the gate, and the distance from the bottom at which it acts.

7. A hydraulic press is to lift a load of 12 tonf. The area of the load piston is 5 ft^2 and the area of the small piston is $\frac{1}{2}$ ft^2. Assuming an efficiency of 100 per cent, calculate (*a*) the effort applied at the small piston to raise the load (*b*) the distance moved by the effort when the load is raised 4 in.

8. A hydraulic press is used to raise a load of 1000 kg. The ram has a diameter of 30 cm and the small piston a diameter of 5 cm. Assuming an efficiency of 100 per cent, calculate (*a*) the effort required at the small piston end to raise the load, and (*b*) the pressure on the face of the small piston.

9. In a hydraulic press the diameter of the large piston is 18 in. and the diameter of the small piston is 4 in. If a load of 2 tonf is to be raised, calculate

(*a*) the pressure on the face of the small piston
(*b*) the effort required to raise the load
(*c*) the work done in raising the load 3 in.
Assume an efficiency of 100 per cent.

10. A submerged rectangular sluice gate has vertical sides of 1 m and horizontal sides of 0·7 m. If the top side of the gate is 1·2 m below the surface of the water calculate the total thrust on the gate.

11. A circular orifice in a vertical wall is used to discharge water from one side. Find the total thrust on a plate 50 in. diameter, covering the orifice, when its centre is 10 ft below the surface of the water. Water weighs 62·5 lbf/ft^3.

12. In a hydraulic press the ram has a diameter of 300 mm and the small piston a diameter of 25 mm. If the pressure in the pipe line is 2,000 kN/m^2, calculate the effort required to produce this pressure, and the load which the ram will raise, if the efficiency of the press is 85 per cent.

13. (*a*) State Archimedes' principle. (*b*) A solid of relative density 2·4 weighs 21 g in water and 30 g in a liquid. Calculate the weight of the solid in air and the relative density of the liquid.

14. (*a*) Describe how the specific gravity of a liquid may be determined experimentally by totally immersing and weighing a solid object first in water and then in the liquid. What "principle" is involved? (*b*) Calculate the specific gravity of an oil if a piece of lead weighing 3 lbf weighs 2·7 lbf when totally immersed in the oil, and 2·6 lbf when totally immersed in water. (N.C.T.E.C.)

15. A cylindrical tank 3 m high is fitted with a vertical outlet pipe 75 mm diameter in its flat base. A spring-loaded disc valve fits over the lower end of the pipe 2·5 m below the base of the tank. If the spring load on the valve is 200 N, determine the depth of water in the tank when the valve is about to open. What increase in spring load is required if the valve opens when the tank is just full? (N.C.T.E.C.)

19

Laboratory experiments

It is now laid down in regulations that students taking a course in Mechanical Engineering Science must devote a stated proportion of their studies to work of an experimental nature, and further, must complete several experiments, and write some of them up fully.

A laboratory experiment affords an opportunity for the student to

(*a*) put theory into practice,

(*b*) familiarize himself with certain types of apparatus and equipment,

(*c*) learn techniques in the use of equipment,

(*d*) learn how to conduct an experiment,

(*e*) learn how to write up results.

In a session of two hours there is no time to waste if sometimes two experiments have to be completed and the results worked out, and so it is necessary to get down to the business of making the experiment without delay.

Some of the apparatus may be expensive and even delicate, so that you must remember that you are approaching a mature age when you must think of what might happen to expensive college apparatus should you rashly turn handles and twiddle knobs before you have even read instructions and become familiar with the apparatus you are about to use.

It is often the practice to hand out a laboratory sheet to every student at the beginning of an experiment. On this sheet is given information concerning the object of the experiment, how it is to be made, and what is required in the way of tabulated results and graphs. You should therefore read through the laboratory sheet first, and then familiarize yourself with the apparatus before making the experiment.

As soon as you have ascertained what you are expected to do, and how the apparatus functions, get to work and begin the experiment. Think of what you are doing, and what the results mean.

Do not just collect a few results hurriedly and then abandon the apparatus to write up what you have done. Satisfy yourself first that you have obtained all the information required, and then check your results with theory or by plotting a graph BEFORE you dismantle the apparatus, or put it away. Identify the apparatus you use by its serial number if possible.

Never anticipate your answer, and never try to obtain readings or answers to fit what YOU imagine to be the correct result. Every reading and measurement should be obtained as accurately as is possible, and written down exactly as obtained.

Writing-up the results

Very few experiments are of such a minor nature that it is not worth while describing either the apparatus used or the method of making the test. In the opinion of the authors every experiment, made at this and at higher levels, should be written up properly—that is, in an approved manner, and in ink. The pattern is a simplified version of the technical report adopted by government establishments and most of our large industrial organizations. The purpose is to put down in a logical form the object of the experiment, how it was made, and the results achieved. Each experiment should have a number, a title and a date.

The work should be written up in the third person and past tense under the following headings:

1. Object.
2. Description of apparatus.
3. Method of test.
4. Results obtained.
5. Theory and discussion.
6. Conclusions.

Where possible, sketches of apparatus and graphs should be added to the written work.

The object of the experiment should be defined clearly. For instance, the title of the experiment may be 'The Simple Pendulum', but the object of the test could be 'To determine the magnitude of the acceleration due to gravity'.

The paragraph headed 'Description of apparatus' should give sufficient detail for the reader to be able to visualize the complete apparatus, whether he is familiar with it or not. Obviously the value of the paragraph would be greatly enhanced if a line diagram, appropriately labelled, were provided. A line diagram means that all unnecessary detail is omitted, such as nuts, washers, bolts, thickness of metal, etc.

A mere list of the apparatus used, including such trivial items as rubber tubing, clips, etc., is not required. Such items, and many others, including thermometers, if considered necessary to mention for the proper functioning of the apparatus could be labelled in the diagram.

All essential detail should be mentioned in the 'Method of Test'. If any precautions or certain procedures are advised, these should be stated clearly, and, where necessary, reasons should be given.

Every result should be written exactly as obtained, even when the experiment is repeated two or three times. The results should be tabulated, and may then be converted, if necessary, into a form more suitable for plotting as graphs. Each graph should be drawn on squared paper and as large as possible. The axes and the scales should be labelled clearly. If more than one curve is plotted on one page—this should be avoided wherever possible—the points should be defined clearly by using different symbols, and a key should be provided. Different coloured inks often make it easier to distinguish curves which are near to each other.

If a theory is involved it should be stated, and the results, including the shape of the graph, should be discussed in relation to the theoretical result. Any errors and discrepancies can also be discussed, but every student should remember that these are more often caused by an inexperienced technician than by faults in the apparatus.

The conclusions must be factual and concise. They must also be related very closely, in the first instance, to the object of the experiment.

In some experiments the apparatus requires very little description, whereas in others some portions of the apparatus require considerable explanation. Sometimes the results can be obtained very quickly, but calculations are necessary to show how the results agree with theory. Clearly, therefore, the amount of work involved in each section of the 'write-up' varies from one experiment to the next.

One experiment, entitled 'The Simple Pendulum' has been written up fully, and may be used by the student as a model.

Experiment No. 1. ***The Simple Pendulum.*** 30th September, 1968

Objects

1. To verify the relationship between the length of the pendulum and its time of oscillation.
2. To determine the magnitude of the acceleration due to gravity.

Description of apparatus

The apparatus consisted of a heavy brass bob attached by a length of thread to a fixed position. This fixed position was obtained by passing the thread between the two halves of a split cork which was held in a clamp on a retort stand (figure 19.1).

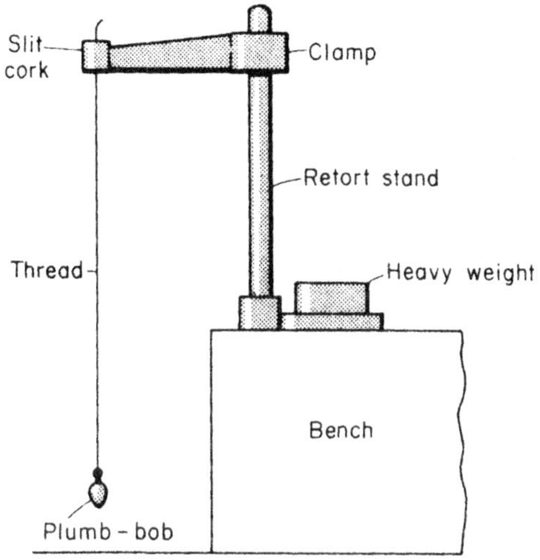

Figure 19.1 The simple pendulum.

Method of test

The length of the thread was measured from the underside of the cork to the centre of the bob by means of a rule. This length was adjusted to 3 feet by drawing the thread through the cork, and the thread was then held firmly by tightening the clamp. The retort stand was placed on the edge of the bench so that the pendulum was free to move sideways.

When the pendulum was hanging vertically and stationary, a chalk mark was made on the bench immediately behind the thread. The pendulum was then caused to swing in one plane with an amplitude not exceeding one sixth of the measured length.

An observer seated himself in front of the apparatus such that the chalk line, the suspension thread and the clamp stand overlapped when the pendulum was in its vertical position. Then, by means of a stop watch, the time was recorded for the pendulum to complete 20 oscillations. A new oscillation was made each time the pendulum crossed the vertical chalk mark whilst moving in the same direction. The time of 20 oscillations was determined 3 times.

This procedure was repeated for different lengths of thread, the length being reduced each time by 6 inches and finally TO 4 inches.

The following precautions were taken:

1. Short lengths of the thread were measured extra carefully.

2. The timekeeper maintained his position during a test so as to avoid parallax.

3. The stop watch was started and stopped as the pendulum crossed the chalk line.

4. If the path of the pendulum became elliptical the test was stopped and restarted.

Results obtained

Time for 20 oscillations (seconds)	*Average time for 20 osc. (second)*	*Time for 1 osc. (T sec)*	T^2	*Length of thread (in.)*	*Length of thread (l ft)*
38·3 38·2 38·3	38·27	1·913	3·66	36	3·0
34·9 34·9 34·9	34·90	1·745	3·05	30	2·5
31·2 31·2 31·2	31·20	1·560	2·43	24	2·0
27·1 27·1 27·0	27·07	1·353	1·83	18	1·5
22·0 22·2 22·1	22·10	1·105	1·22	12	1·0
15·8 15·8 15·7	15·77	0·789	0·62	6	0·5
12·8 12·8 12·7	12·77	0·638	0·41	4	0·33

Graph

A graph was plotted (figure 19.2) using the value of l as ordinates and T^2 as abscissae. The best straight line was drawn through the points and its slope was determined.

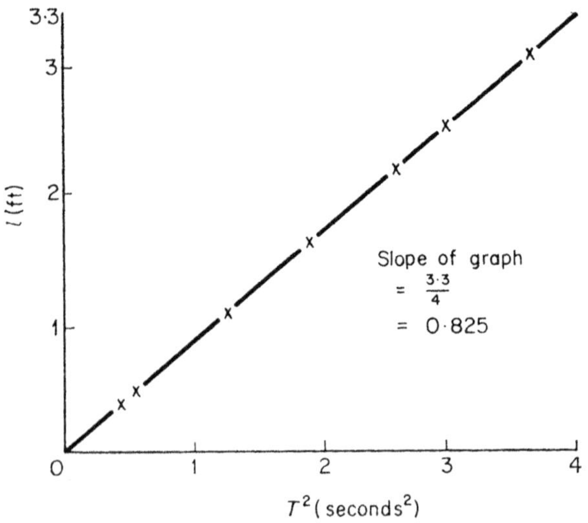

Figure 19.2 Graph of $l - T^2$.

The value of the slope obtained was

$$\frac{l}{T^2} = 0{\cdot}825$$

Theory and discussion
For a simple pendulum the time of oscillation is given by

$$T = 2\pi\sqrt{\frac{l}{g}}\ \text{seconds}$$

Squaring and re-arranging

$$\begin{aligned} g &= 4\pi^2\left(\frac{l}{T^2}\right) \\ &= 39{\cdot}48 \times 0{\cdot}825 \\ &= 32{\cdot}5\ \text{ft/s}^2 \end{aligned}$$

(The value should be $32{\cdot}2$ ft/s^2)

At short lengths of the thread an error of one tenth of an inch on four inches represents an error of $2\frac{1}{2}$ per cent in measurement. Because g is directly proportional to l the value of g might consequently vary—if determined at this length—from $31{\cdot}4$ to $33{\cdot}0$ ft/s^2.

The average time of one oscillation should not be in error by more than 1 per cent. However, because g is inversely proportional to T^2 the value of g can consequently be in error by 2 per cent from this cause.

The biggest source of error appears to be in estimating the centre of gravity of the plumb bob, as this affects an accurate determination of l. Such an error should make itself apparent on the graph.

Conclusions

1. There is a linear relationship between l and T^2.
2. The value of g obtained was $32{\cdot}5$ ft/s^2.

Answers to additional exercises

Chapter 2, p. 29

1. 1·6 m/s^2.
2. $\frac{5}{24}$ ft/s^2; 2 min 56 s.
3. (*a*) 1·20 ft/s^2, (*b*) 20·3 miles.
4. (*a*) 80 km/h, (*b*) 90 s, (*c*) $\frac{16}{81}$ m/s^2.
5. (*a*) 14·6 ft/s^2, (*b*) 7·02 s.
6. 6 m/s.
7. (*a*) 100 ft, (*b*) 2·78 s, (*c*) 89·2 ft/s.
8. (*a*) 31·8 m, (*b*) 1·93 s, (*c*) 5·1 s.
9. (*a*) 1·26 × 10^4 ft, (*b*) 9·1 ft/s.
10. 2·47 m/s.
11. (*a*) 22$\frac{1}{3}$ rad/s^2, (*b*) 400 rev.
12. (*a*) $\frac{7}{90}$ rad/s, (*b*) 893 rev, (*c*) 26·2 m/s.
13. (*a*) 25$\frac{1}{2}$°N of E, (*b*) 38$\frac{1}{2}$ mile.
14. 59° 36′ E of N; 42 knots.
15. 28·6 km/h at 36° 20′ N of E, (*b*) 2·27 km.
16. (*a*) 47·4 rad/s, (*b*) Rim top 102·6 ft/s in direction of motion, rim bottom 0 ft/s, rim sides 72 ft/s at 45° to verticals.

Chapter 3, p. 43

1. (*a*) 32·7 s, (*b*) 0·934 m/s^2, (*c*) 934 N.
2. 24·5 kN.
3. 28 lbf/tonf.
4. 3·157 t.
5. (*a*) 2415 lbf, (*b*) 2240 lbf, (*c*) 2065 lbf.
6. 26·7 N.

Chapter 4, p. 48

1. 49·05 kJ; 5·886 kJ; 43·16 kJ.
2. 39·2 m/s, 156·9 kJ, 156·9 kJ.
3. 3750 ft lbf.
4. 312·9 kJ, 251·2 kJ.
5. 144,900 ft lbf.
6. (*a*) 61·8 kJ, (*b*) 1·39 kJ.
7. 1·84 tonf; 17,015 ft tonf; 17,015 ft tonf.

Chapter 5, p. 56

1. 268,900 ft lbf.
2. (*a*) 740·6 kJ, (*b*) 65·83 kJ.
3. 444 ft/s^2.
4. 4033 ft/s^2.
5. 360 lbf.
6. 19·4 m/s.
7. 8·152 kN.
8. 2230 lbf.
9. (*a*) 1003 lbf, (*b*) 29·9 mile/h.
10. 26·5 kN; 12·73 kN.
11. 40·02 N.
12. 895 lbf.

Chapter 6, p. 83

1. (*a*) 0·6 kN; (*b*) 0·59 kN.
2. (*a*) 58·86 N; (*b*) 612·14 N.
3. (*a*) 27·7 lbf; (*b*) 0·9 lbf.
4. (*a*) 0·239; (*b*) 13$\frac{1}{2}$°.
5. 0·16.
6. 127·5 N.
7. (*a*) 0·33; (*b*) 18° 18′.
8. 253·1 N; 17°.
9. 134 lbf.

Chapter 7, p. 125

1. 1·65 kN; 72·7 per cent.
2. $P = 83{\cdot}4 + 3{\cdot}924$ N.
3. 115 lbf.
4. 24·08 kN.
5. (*a*) 1715 lbf; (*b*) 714 lbf.
6. (*a*) 5; (*b*) 1·496 kN; (*c*) 4.
7. (*a*) 80 per cent; (*b*) 10 lbf; (*c*) 50 ft lbf.
8. (*a*) 238 lbf; (*b*) 95·3 lbf.
9. 981 N.
10. (*a*) 112 lbf ft; (*b*) 4·27.
11. (*a*) 22·4 lbf ft; (*b*) 1·07.
12. 1375 lbf; 550 lbf.
13. 75·6 lbf; 25·2 lbf.
14. 20·9 per cent.
15. 39·6 per cent.
16. 1·4 mm.
17. 51·4 lbf; 669 ft lbf.

Chapter 8, p. 152

1. 5·2 lbf horizontally; 40·3 lbf at $82\frac{1}{2}°$ to the horizontal.
2. 34·3 N horizontally; 237·4 N at $81\frac{1}{2}°$ to the horizontal.
3. (*a*) 15 lbf, (*b*) 30 lbf at 18° to the vertical.
4. (*a*) 1·349 kN, (*b*) 182·5 kgf at $49\frac{1}{2}°$ to the vertical.
5. (*a*) 78 lbf, (*b*) 78 lbf at 50° to the vertical.

6.

Member	*Force-kN*	*Type*	*Member*	*Force-kN*	*Type*
Reaction, A	18·39	—	DE	17·17	Strut
AC	36·78	Strut	EF	14·71	Strut
AF	31·88	Tie	EB	31·88	Strut
CD	17·17	Strut	BF	27·96	Tie
CF	19·62	Strut	Reaction, B	15·94	—
DF	17·17	Tie			

7.

Member	*Force-tonf*	*Type*	*Member*	*Force-tonf*	*Type*
Reaction, A	2·75	—	DE	6·5	Strut
AC	5·5	Strut	DG	2·7	Tie
AF	4·75	Tie	EB	6·5	Strut
CD	5·5	Strut	EG	2·0	Strut
CF	1·0	Strut	BG	5·65	Tie
DF	1·35	Tie	Reaction, B	3·25	—
FG	3·9	Tie			

8.

Member	*Force-kN*	*Type*	*Member*	*Force-kN*	*Type*
Reaction, A	38·26	—	DE	1·177	Strut
AC	53·95	Strut	DB	56·7	Strut
CE	1·47	Tie	BE	40·22	Tie
AE	38·45	Tie	Reaction, B	40·22	—
CD	40·22	Strut			

9.

Member	*Force-tonf*	*Type*	*Member*	*Force-tonf*	*Type*
Reaction, A	1·375	—	DE	0·875	Tie
AC	2·75	Strut	DB	3·0	Strut
AE	2·375	Tie	BE	1·5	Tie
CE	0·875	Strut	Reaction, B	2·625	—
CD	2·25	Strut			

10.

Member	*Force-kN*	*Type*	*Member*	*Force-kN*	*Type*
Reaction, A	294·3	Downwards	CD	294·3	Tie
AC	277·5	Tie	DB	131·4	Strut
AB	219·7	Strut	DE	262·9	Tie
BC	219·7	Strut	BE	248·2	Strut
Reaction, B	490·5	Upwards			

Chapter 9, p. 167

1. (*a*) 10·18 tonf/in^2, (*b*) 0·0076 in.
2. (*a*) 16 $\times$ 10^7 N/m^2; 24·96 $\times$ 10^7 N/m^2, (*b*) 0·249 cm.
3. 275 lbf; 0·092 in.
4. (*a*) 12·6 $\times$ 10^7 N/m^2, (*b*) 6 $\times$ 10^{-4}, (*c*) 209·9 $\times$ 10^9 N/m^2.
5. 3·47 tonf/in^2.
6. 28·72 $\times$ 10^7 N/m^2; 0·00681 cm.
7. 10·6 tonf.
8. 43·75 $\times$ 10^7 N/m^2; 14·58 $\times$ 10^7 N/m^2; 71·51 kN.
9. 41·2 tonf; 68·6 tonf/in^2.
10. 0·102 cm.

Chapter, 10 p. 193

1.

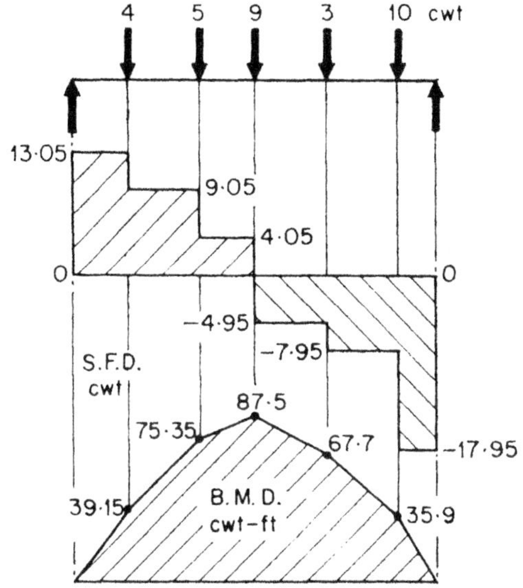

2.

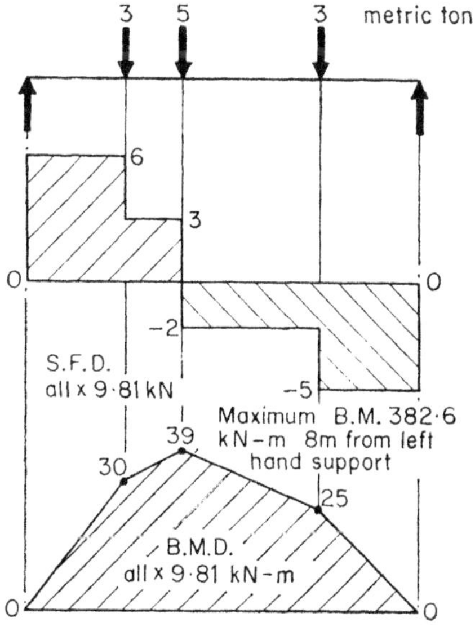

Chapter 10, p. 194

3.

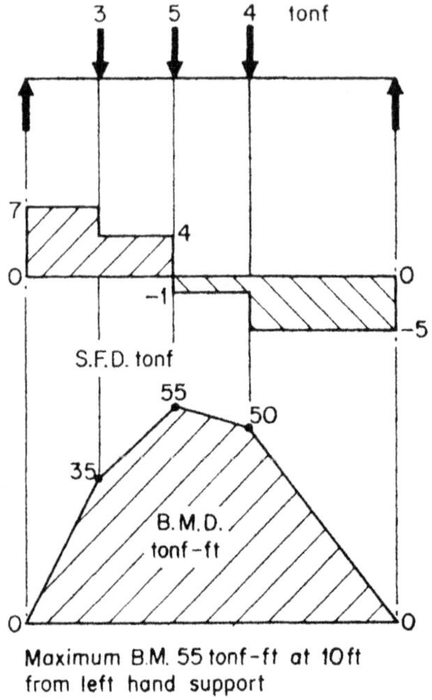

4.

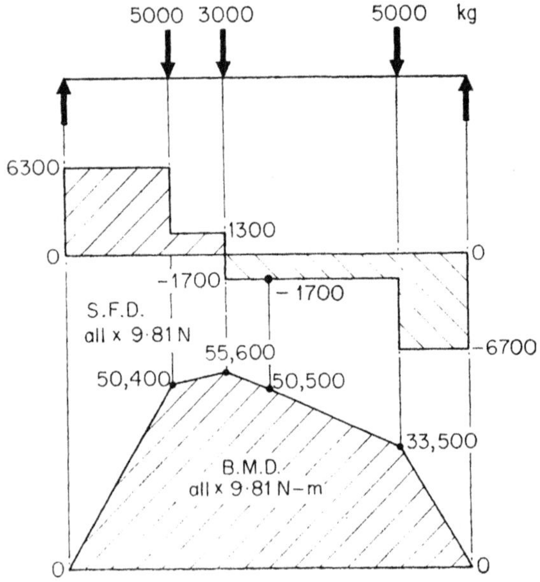

Chapter 10, p. 194

5.

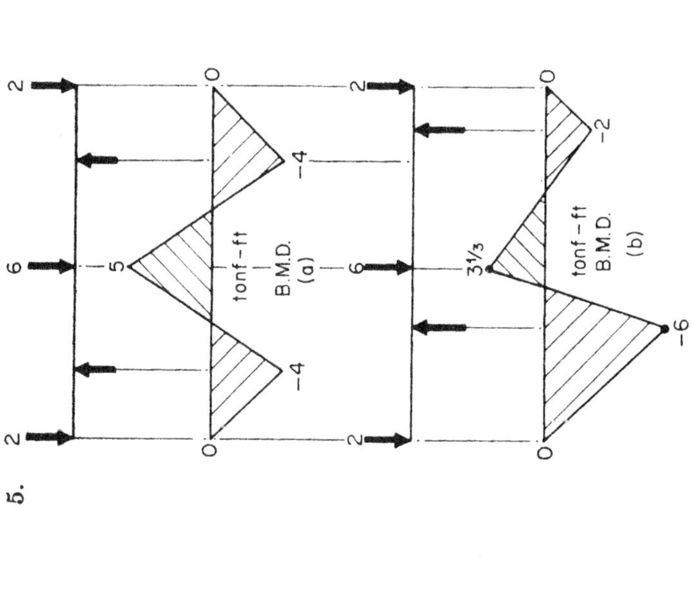

6.

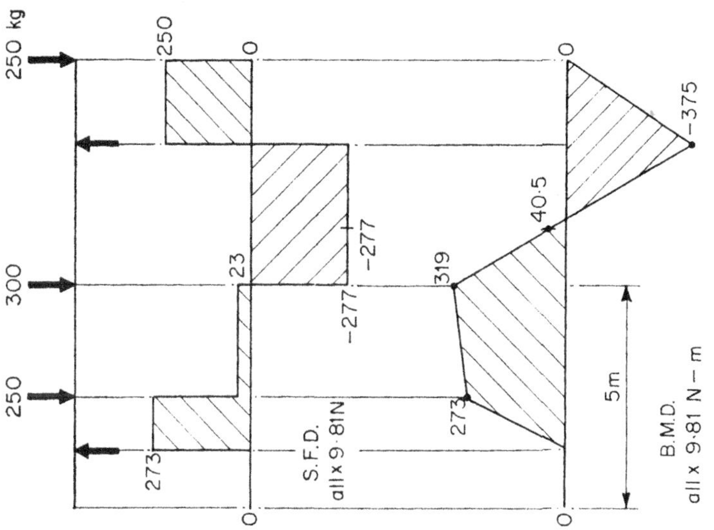

Chapter 10, p. 194

7.

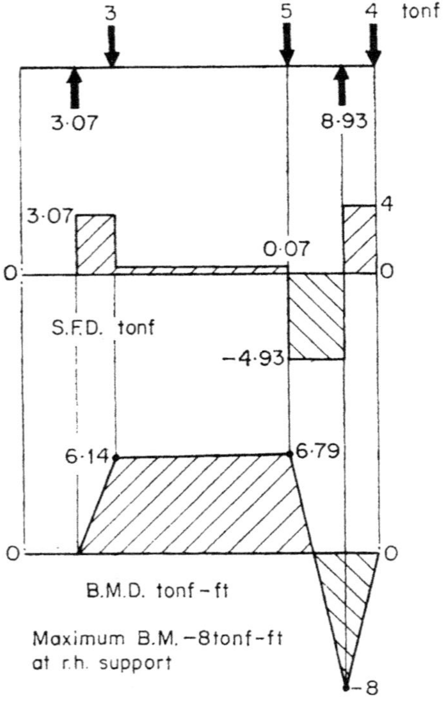

Maximum B.M. −8tonf−ft
at r.h. support

8.

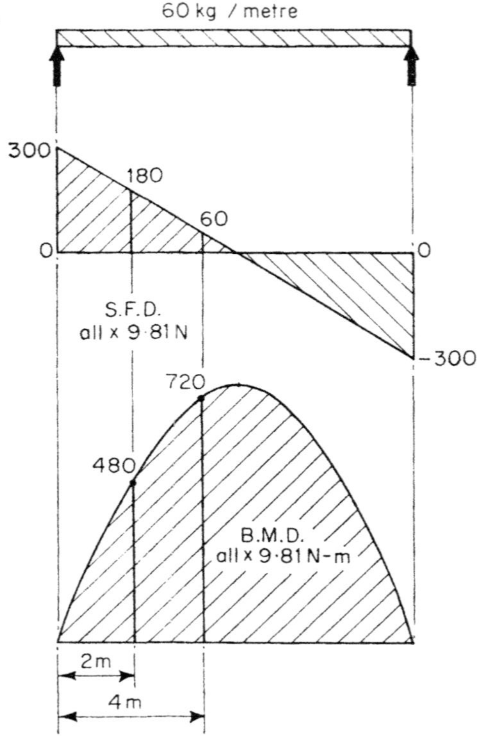

Chapter 10, p. 194

9.

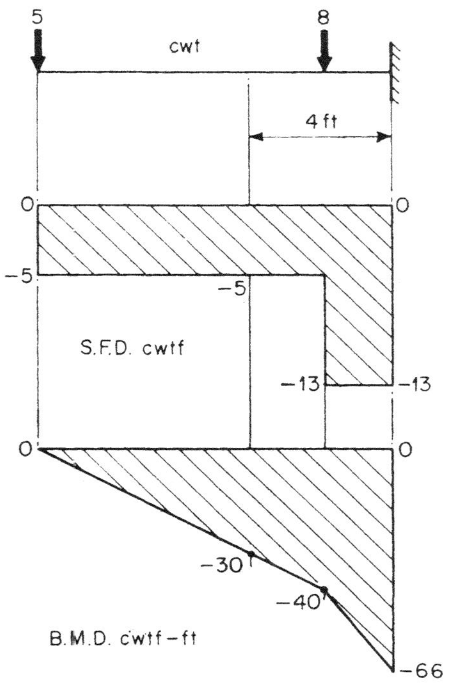

(a) S.F. −13cwtf B.M. −66cwtf−ft
(b) S.F. −5cwtf B.M. −30cwtf−ft

10.

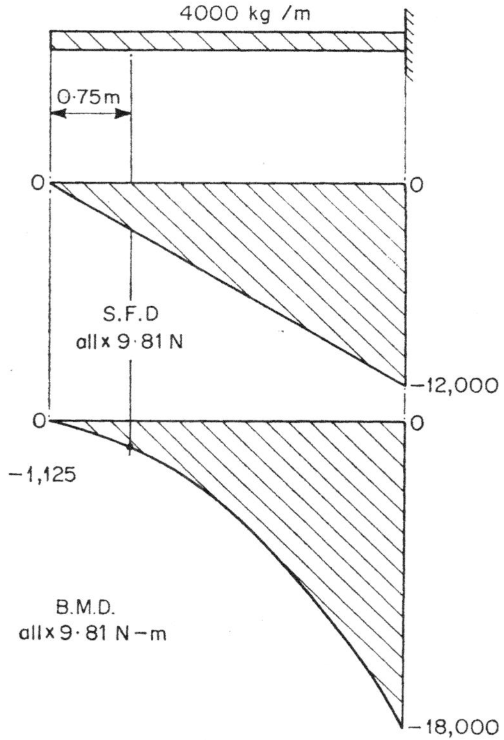

Chapter 11, p. 206

1. 39·9 in. from bottom edge; 31·5 in. from right hand edge.
2. 0·81 cm to the left; 0·18 cm down.
3. 0·143 in. to the left of the centre.
4. 7·90 cm from bottom edge; 8·31 cm from right hand edge.
5. 6·475 in.
6. 0·333 cm.
7. 8·25 in. from free end of cylinder (*a*).
8. 23·5 cm from end of cylinder.
9. 2·22 in.
10. 1·58 metre from either end of bar, on internal bisector of right angle.
11. (*a*) 3·40 in., (*b*) 4·28 in.
12. 6·88 cm above bottom of section.

Chapter 12, p. 222

1. 158°F; 50°C.
2. 0·2 degF low.
3. −40°C −40°F.
4. 313°C.
5. 0·26 m^3.
6. 322 degF decrease.
7. −143°C.
8. 90·5 lbf.
9. 12.
10. 10·24 min.
11. 12·73 b.
12. (*a*) 96 ft lbf/lbf degK, (*b*) 53·3 ft lbf/lbf degR.
13. 0·62 m^3; 0·849 m^3.

Chapter 13, p. 233

1. 13 kg.
2. 61·44 kJ/kg.
3. 0·965.
4. (*a*) 2715 kJ/kg, (*b*) 2648 kJ/kg.
5. 725 Btu/lbf.
6. 1005 kJ.
7. (*a*) 2567·2 Chu; (*b*) 2170·4 Chu; (*c*) 2790·2 Chu.
8. (*a*) 5875 Btu; (*b*) 5321·5 Btu; (*c*) 6100 Btu.
9. 0·787.
10. (*a*) 1080·6 Btu/lbf; (*b*) 1167·1 Btu/lbf; (*c*) 1187·9 Btu/lbf.

Chapter 14, p. 242

1. 64·8 per cent.
2. 76·4 per cent.
3. 10,260 lbf/min.
4. 1337·1 litre/min.
5. 53 per cent; 0·19 m^2/kg.

Chapter 16, p. 254

1. (*a*) 10·8; (*b*) 8·1; (*c*) 75 per cent.
2. (*a*) 24·3 per cent; (*b*) 18·2 per cent.
3. 34·4 per cent.
4. (*a*) 31·1 per cent; (*b*) 37·9 per cent.
5. (*a*) 8·29; (*b*) 10·15; (*c*) 81·6 per cent; (*d*) 21·1 per cent; (*e*) 25·9 per cent.
6. (*a*) 20·5 l/min; (*b*) 25·8 kW.

Chapter 17, p. 265

1. 11·46 kg.
2. (*a*) CO_2 25·9 per cent; N_2 74·1 per cent; (*b*) CO_2 25·0 per cent; N_2 71·4 per cent; H_2O 3·6 per cent.
3. (*a*) 12·3 lbf; (*b*) CO_2 20·0 per cent; N_2 75·9 per cent; O_2 4·1 per cent.
4. 15·5 kg.
5. 10·95 lbf.
6. CO_2 15·8 per cent; H_2O 3·9 per cent; N_2 73·0 per cent; Excess O_2 7·3 per cent.
7. (*a*) 11·09 kg; (*b*) CO_2 16·8 per cent; O_2 7·0 per cent; N_2 76·2 per cent.
8. CO_2 10·65 per cent; O_2 7·22 per cent; N_2 82·13 per cent.
9. (*a*) 11·86 lbf; (*b*) 26·9 per cent.

Chapter 18, p. 280

1. 514 tonf.
2. 73·5 kN.
3. (*a*) 437·5 lbf/ft^2; (*b*) 218·75 lbf/ft^2; (*c*) 244 tonf; (*d*) 17·1 tonf; (*e*) 34·2 tonf.
4. (*a*) 24·1 tonf; 2·68 tonf; (*b*) 21·42 tonf at 4·33 ft from bottom.
5. 490·5 kN at 2·53 m from bottom.
6. 493·5 tonf at 10·31 ft from bottom.
7. (*a*) 1·2 tonf; (*b*) 40 in.
8. (*a*) 272·5 N; (*b*) 13·85 $\times$ 10^4 Newtons/m^2.
9. (*a*) 17·6 lbf/in^2; (*b*) 221·5 lbf; (*c*) 0·5 ft tonf.
10. 1190 kg.
11. 548 tonf.
12. 982 N; 120·2 kN.
13. 36 g; 0·4.
14. 0·75.
15. 2·115 m; 25 N.

Index

Printed in Dunstable, United Kingdom

91898072R00178